T0292595

Studies in Systems, Decision and Control

Volume 39

Series editor

Janusz Kacprzyk, Polish Academy of Sciences, Warsaw, Poland
e-mail: kacprzyk@ibspan.waw.pl

About this Series

The series "Studies in Systems, Decision and Control" (SSDC) covers both new developments and advances, as well as the state of the art, in the various areas of broadly perceived systems, decision making and control- quickly, up to date and with a high quality. The intent is to cover the theory, applications, and perspectives on the state of the art and future developments relevant to systems, decision making, control, complex processes and related areas, as embedded in the fields of engineering, computer science, physics, economics, social and life sciences, as well as the paradigms and methodologies behind them. The series contains monographs, textbooks, lecture notes and edited volumes in systems, decision making and control spanning the areas of Cyber-Physical Systems, Autonomous Systems, Sensor Networks, Control Systems, Energy Systems, Automotive Systems, Biological Systems, Vehicular Networking and Connected Vehicles, Aerospace Systems, Automation, Manufacturing, Smart Grids, Nonlinear Systems, Power Systems, Robotics, Social Systems, Economic Systems and other. Of particular value to both the contributors and the readership are the short publication timeframe and the world-wide distribution and exposure which enable both a wide and rapid dissemination of research output.

More information about this series at http://www.springer.com/series/13304

Robert Kozma · Walter J. Freeman

Cognitive Phase Transitions in the Cerebral Cortex - Enhancing the Neuron Doctrine by Modeling Neural Fields

With Commentaries by: Kazuyuki Aihara and Timothy Leleu, Bernard Baars, Steven Bressler, Ray Brown and Morris Hirsch, Péter Érdi and Zoltán Somogyvári, Hans Liljenström, Frank Ohl, Ichiro Tsuda, Giuseppe Vitiello, Paul Werbos, and James Wright

Robert Kozma
Department of Mathematical Sciences
University of Memphis
Memphis, TN
USA

Walter J. Freeman
Division of Neurobiology
University of California at Berkeley
Berkeley, CA
USA

ISSN 2198-4182 ISSN 2198-4190 (electronic)
Studies in Systems, Decision and Control
ISBN 978-3-319-24404-4 ISBN 978-3-319-24406-8 (eBook)
DOI 10.1007/978-3-319-24406-8

Library of Congress Control Number: 2015950877

Springer Cham Heidelberg New York Dordrecht London

Printed on acid-free paper

Springer International Publishing AG Switzerland is part of Springer Science+Business Media
(www.springer.com)

This work is dedicated to scientists who persevere in questioning prevailing dogma in search of wisdom in the frontiers of neuroscience

Preface

Everyday subjective experience of the stream of consciousness suggests continuous cognitive processing in time. Brain monitoring techniques with markedly improved spatiotemporal resolution, however, provide an evidence of discontinuities between transiently stationary dynamics in brains. We observe spatiotemporal cortical dynamics as giving sequences of metastable spatial patterns of coherent population activity. Each metastable pattern manifests a cortical state (corresponding to a cinematic frame) that collapses in transient desynchronization (analogous to a shutter), followed by the rapid emergence of a new pattern. The temporal sequence of metastable spatial patterns is closely correlated with intentional behaviors. Each frame manifests the action–perception cycle [1] by which animals probe their environments and learn about them by accommodating to the impact of their own actions on their sensory systems.

Patterns of microscopic pulse trains from depth recordings, mesoscopic action potentials from intracranial assemblies, macroscopic surface electrocorticograms (ECoG), scalp electroencephalograms (EEG) and magnetoencephalograms (MEG), and high-resolution functional magnetic resonance images (fMRI) reveal a hierarchy of brain states across temporal and spatial scales. The observed neural processes have significant high-frequency spatiotemporal components. The high-frequency oscillatory components measured by brain monitoring techniques are widely viewed as noise and/or artifacts and eliminated by averaging and band-pass filtering. However, the seemingly erratic dynamic behavior of neural field potentials contains recognizable patterns, which have been measured after appropriate decomposition into wavelets [2].

Analysis of high-frequency evoked potentials measured by high-resolution brain imaging techniques point to frequent transitions between periods of large-scale synchronization and intermittent desynchronization at alpha-theta rates. These observations support the hypothesis about the cinematic model of cognitive processing, according to which higher cognition can be viewed as multiple movies superimposed in time and space. The metastable spatial patterns of field potentials manifest the frames, and the rapid transitions provide the shutter from each pattern to the next in multiple streams of various sizes.

The alternating states offer a stark contrast in organization of activity between a low-energy, sparsely firing state that we analogize to a gas-like phase and a high-energy liquid-like phase, still sparsely firing but synchronized. We analogize the onset as a phase transition of condensation, followed by evaporation after 3 to 5 cycles of the oscillation [3]. The cortex holds itself in a state of criticality, which is manifested in a readiness to reorganize itself from random noise into synchronization of immense populations. The synchrony may not be apparent in single cell recordings, partly because the cortical neurons are time-multiplexing to distribute the computational load over a large number of participants, and partly because many of the neurons in a pattern are being held silent, told to shut up, because every pattern requires both light and dark.

What this analysis tells us is that the firing of pulses by axons (and by some dendrites) is only a part of the story being told by brain activity. The other part is told by the flows of ionic current from dendrites that determine the firing. The value of this other part is enhanced by the fact that the discovery of large-scale synchronization–desynchronization transitions in brains open new opportunities to the development of brain computer interfaces (BCIs). In clinical settings, BCIs can help to diagnose, predict, and treat cognitive diseases at the early stage; they can also drastically improve the quality of life of disabled people. The potential benefits are enormous. The implementations can be divided into two broad groups. Invasive techniques involve the placement of implants on brains by opening the skull. The implants can have single electrodes or multiple arrays. Rapid technological development allows to access information from individual neurons and to achieve the goal of breaking the neural code of the brain to be able to zero in on individual neurons [4, 5]. But there is no one code, if there is any at all, and if one is to be accepted for axons, another must be accepted for dendrites.

In noninvasive devices, the electrodes are located on the scalp far from the cortical neurons. As a result, the recorded signals and images lack the high resolution observed in invasive devices and cannot give us an axonal code. Noninvasive approaches give us access to dendritic codes, as they are applicable in everyday life. As a result, noninvasive BCIs are increasingly accessible, for example, in the entertainment industry, as well as serving as personal assistants in physical training and exercises.

BCIs are young and immature technologies, and they are still at the very early stage of their development. In spite of the advances with noninvasive approaches, extracting meaningful information from the signals of remotely located electrodes may seem daunting. Indeed, significant exponents of the neuroscience community consider the obstacles impenetrable and the related activities outright meaningless. The task of noninvasive brain monitoring has been ridiculed by comparing it to the assumed "impossibility" of the assignment of Keystone Cops to eavesdrop on a single conservation in a stadium from outside:

Keystone Cops in a crowded stadium, illustrating the alleged paradox of brain monitoring using noninvasive devices (Illustration by Vladimir Taytslin)

> External devices, such as the brainwave-reading skull cap … marketed as "having applications for wellness, education and entertainment," have none of these risks. But because their sensors are so far removed from individual neurons, they are also far less effective. They are like Keystone Cops trying to eavesdrop on a single conversation from outside a giant football stadium. [4].

This misguided simile betrays the lack of awareness of large-scale brain dynamics. To pursue the simile, the *roar of the crowd at a football game* is not the sum of many thousands of conversations. It is the collective action of spectators engaged in a social ritual, for which the stadium was built at enormous expense. The crowd has convened with extensive planning in advance to enjoy participation in the realization of social solidarity. Comparably, millions of neurons form collectives that transcend pairwise synaptic exchanges. The collective that they form has the power of numbers in synchronized discharges sweeping through the basal ganglia and brain stem. The firings are far better positioned and organized than the spike trains of networks of a few tens of neurons in an unknown number of networks widely spaced in cortex. Undeniably there must be private conversations among neurons, but they are not the whole story. Collective actions take precedence, and these are observable without need for neurosurgeons.

The alternation between synchronized and desynchronized states is by no means obvious in casual recordings of ECoG or scalp EEG. Some form of temporal filtering is required [2, 6] used the spatial gradient of the alpha amplitude in the scalp EEG. Brockmeier [7] used ICA; Ruiz et al. [8] used the Rician statistics to specify a band of beta activity. Zhang et al. [9] used the phase coherence in the beta and gamma bands. Panagiotides et al. [10] used the spatial standard deviation of the beta amplitudes as a marker. Each of these measures gave access to the AM patterns that provided the neural correlates of perception of conditioned stimuli in the several modalities. The techniques also served to show that the low-energy gas-like supervenes in a resting state that was observed in animals and humans when they are placed in monotonous and unchanging environment that induces awake rest. This enabled us to define the resting state seen also under light anesthesia as a sustained ground state with a simple $1/f^{\alpha}$ canonical EEG/ECoG temporal spectrum with no peaks in the beta or gamma ranges, and $2 \leq \alpha \leq 4$ [11].

The neurodynamics of the resting and active states thus defined have been tested for stationarity and linearity. This was done by perturbing the cortex with electric impulses modeled with the Dirac delta function and measuring the impulse responses by fitting them with wavelets consisting of sums of linear basis functions [12]. The basis functions yielded the characteristic frequencies of the cortices in their normal operating range, which was defined by the range of ± 3 standard deviations of the amplitude of the resting or working EEG/ECoG. Compliance with superposition in the small signal range thus defined made it possible to model the cortical dynamics with the solutions to ordinary differential equations (ODEs). That in turn made it feasible to model the strengths and signs of synaptic couplings with adaptive coefficients and to replace the nonlinear gain curve with the slope of the tangent to the curve at the estimated operating point.

By these steps it became possible to model the dynamics on both sides of the phase transition, the random ground phase and the high-energy active phase. However, the ODE could not model the phase transition between them. Two approaches have been adopted. The more speculative approach is to use the continuous equations of many-body physics to model the interactions of neural populations that are sufficiently large to enable us to define activity density functions for axonal pulses and postsynaptic potentials (EPSPs and IPSPs) [13].

The more advanced approach is to use Random Graph Theory (RGT) [14, 15] to devise a discrete calculus in which the element is not a neuron but a functional element corresponding to the collective of neurons that participate in time-sharing. Both approaches depend on the basic assumption that the neuropil has the property to sustain pulse trains from many if not most of its neurons but also the property of ephapsis, which operates in a continuum across the sustaining neural population. Ephasis can be modeled by the discrete particles required for digital approximations in both ODE and RGT. The aim of this book is to establish a branch of RGT that supplements and may eventually replace ODE with *neuropercolation* as the basis for modeling neural population dynamics on digital platforms.

In the corresponding mathematical theories, brains are perceived as open thermodynamic systems [3, 16] converting broadly fluctuating sensory data into meaningful knowledge. Among the wide range of approaches addressing discontinuities in brain dynamics, random graphs have unique advantages by characterizing cortical processes as phase transitions and transient percolation processes in probabilistic cellular automata (PCA). The corresponding model is called neuropercolation, which uses results of RGT as a rigorous mathematical approach to formulate the fundamental relationship between transient neural processes in the cortex and the structure of the embedding brain graph. RGT has distinct advantages as compared to differential equations when describing discontinuities in cortical dynamics. It presents a paradigm shift from modeling of individual neurons to modeling the collective behavior of neural populations.

The caricatures provide metaphors to illustrate this paradigm shift and expose the limitations of analytic tools for describing microscopic pulse logic and macroscopic wave dynamics. Just as it is impossible to understand or even conceive the collective dynamics of the football stadium by trying to listen to the individual conversations in the audience; it is a feeble attempt to gain insight into brain dynamics by limiting the scope to individual neurons. Brains are large-scale systems, in which the components produce field effects as emergent phenomenon. It is the fields that provide promise to monitor and understand brain dynamics by attempting to take it apart and then learn how the neurons generate the populations and in turn how the neurons are influenced and controlled by the populations in circular causality.

Our premise is that the repetitive sudden transitions observed in the cortex are maintained by neural percolation processes in the brain as a large-scale random graph near criticality, which is self-organized in collective neural populations formed by synaptic activity. Neuropercolation addresses the complementary aspects of neocortex, manifesting complex information processing in microscopic networks of specialized spatial modules, and developing macroscopic patterns evidencing that brains are holistic, multi-tasking organs. The present volume reviews neurophysiological evidences of collective brain dynamics and proposes neuropercolation as a mathematical model to interpret experimental findings. Potential benefits to brain computer interfaces are indicated, as well.

We are delighted to present this book, which was born out of the many discussions we had in the past 10 years about the role of scale-free structure and dynamics in producing intelligent behavior in brains. The discussions started to converge during the Spring 2006 of Robert's sabbatical visit at Walter's Lab at UC Berkeley. Clearly, the question of scale-free structure and behavior is a controversial and a very contentious issue in the literature. This controversy was apparent in the failed attempts to publish such results in the journal of Behavioral and Brain Science first in 2007 by Walter, then in 2014 as a joint endeavor by two of us. It became clear that different research groups had their vested interests in one or another aspects of the issue and were not interested in hearing or considering alternative points of views. As a result, we have received feedbacks, which in our judgment transcended the boundaries expected in civilized and scientifically solid and justified constructive debates.

Following extensive discussions with our colleagues, we decided to produce this volume, which has a somewhat unorthodox structure. The first half (Part I and II) summarizes our views on the relevant experimental and theoretical findings and methodological issues on intermittent spatiotemporal neurodynamics in the brain and the key role of large-scale, collective oscillations in producing higher cognition and consciousness. The second half of the book (Part III, IV, and V) includes commentaries by leading experts in the field of neuroscience, cognitive science, and theoretical/mathematical modeling of the relationship between microscopic neural level (neuron doctrine) and macroscopic behavioral level (field theories) of brain operation.

We greatly appreciate those who supported our endeavor by contributing to this volume with their commentaries, Kazuyuki Aihara and Timothy Leleu, Bernard Baars, Steven Bressler, Ray Brown and Morris Hirsch, Peter Erdi and Zoltan Somogyvari, Hans Liljenstrom, Frank Ohl, Ichiro Tsuda, Giuseppe Vitiello, Paul Werbos, and James Wright. With their help we are able to present a broad range of views, extending beyond our own constraints and helping to stimulate productive discussions and further breakthroughs in understanding the codes of the brain.

We are thankful for the helpful comments and critical insight by Scott Kelso and Jose Principe, and for encouragement and support from our Springer editors Janusz Kaczprzyk and Thomas Ditzinger.

This volume could not be realized without the support during the past decade from so many of our colleagues, collaborators, mentors, and students, including Paul Balister, Bela Bollobas, Mike Breakspear, A. Brockmeier, Gyuri Buzsaki, Tian Yu Cao, Antonio Capolupo, Jim Caulfield, Joshua Davis, Toshi Fukuda, Grant Gillett, Derek Harter, Mark Holmes, Sanqing Hu, Terry Huntsberger, Roman Ilin, Guang Li, C.T. Lin, Roberto Livi, Vinod Menon, Mark Myers, Masashi Obinata, Sean O'Nuallain, Heracles Panagiotides, Leonid Perlovsky, Sue Pockett, Karl Pribram, Marko Puljic, Rodrigo Quian-Quirga, Misha Rabinovich, Ceon Ramon, Oliver Riordan, Jose Rodriguez, Yusely Ruiz, Miklos Ruszinko, Hava Siegelmann, Rodrigo Silva, Yury Sokolov, Eddie Tunstel, Jun Wang, Anne Warlamount, Ludmilla Werbos, Jian Zhai, and many many more. The excellent drawings by Vladimir Taytslin and Chris Gralapp of EyeArt provide very compelling perspectives to illustrate our messages.

The intended audience of this book includes researchers, postdocs, and graduate students working towards novel approaches in brain science in order to better understand the operation of this very precious and delicate organ we are all equipped with. The results can be useful to maintain normal operation of our brain, help to improve the quality of life of the elderly, and develop novel treatments and approaches for people with cognitive or mental disabilities. In short, help to fulfill the human potential to the highest level and to support the sustainable development of humanity.

Acknowledgments

This work has been supported in part by NSF CRCNS Program DMS-13-11165, DARPA Physical Intelligence Program through HRL subcontract, by AFOSR Lab Task in the Mathematics and Cognition Program to RK, and by grants from the National Institute of Mental Health NIMH (MH06686) to WJF.

Memphis
Berkeley
November 2014

Robert Kozma
Walter J. Freeman

References

1. Merleau-Ponty M (1945) Phnomnologie de la perception. Gallimard, Paris
2. Freeman WJ, Quian Quiroga R (2013) Imaging brain function with EEG: advanced temporal and spatial analysis of electroencephalographic and electrocorticographic signals. Springer, New York
3. Capolupo A, Freeman WJ, Vitiello G (2013) Dissipation of 'dark energy' by cortex in knowledge retrieval. Phys Life Rev, Online. doi:10.1016/j.plrev.2013.01.001
4. Marcus G, Koch C (2014) The future of brain implants. The Wall Street Journal. Dow Jones and Co, S. Brunswick, NJ
5. Sanchez JC, Principe JC (2007). Brain-machine interface engineering (Vol 17). Morgan and Claypool Publishers, LaPorte, Colorado
6. Lehmann WK, Strik B, Henggeler T, Koenig M, Koukkou (1998) Brain electric microstates and momentary conscious mind states as building blocks of spontaneous thinking, I. Visual imagery and abstract thoughts. Int J Psychophysiol 29:111
7. Brockmeier AJ, Hazrati MK, Freeman WJ, Li L, Prncipe JC (2012) Locating spatial patterns of waveforms during sensory perception in scalp EEG. In Engineering in Medicine and Biology Society (EMBC), 2012 Annual International Conference of the IEEE (pp. 2531–2534)
8. Ruiz Y, Pockett S, Freeman WJ, Gonzales E, Guang Li (2010) A method to study global spatial patterns related to sensory perception in scalp EEG. J Neurosci Methods 191:110–118
9. Zhang T, Dai L, Freeman WJ, Li G (2013) EEG spatiotemporal pattern classification of the stimuli on different fingers. In: 4th International Conference on Cognitive Neurodynamics (ICCN2013), Sweden, Springer, Germany
10. Panagiotides H, Freeman WJ, Holmes M, Pantazis D (2008) Behavioral states exhibit distinct spatial EEG patterns. In: Proceedings of the 62nd Annual Meeting, American Epilepsy Society, Seattle, WA, 2008
11. Freeman WJ, Zhai J (2009) Simulated power spectral density (PSD) of background electrocorticogram (ECoG). Cogn Neurodyn 3(1):97–103
12. Freeman WJ (1975/2004) Mass action in the nervous system. Academic, New York. Electronic version 2004. http://sulcus.berkeley.edu/MANSWWW/MANSWWW.html
13. Freeman WJ, Livi R, Obinata M, Vitiello G (2012) Cortical phase transitions, nonequilibrium thermodynamics and time-dependent Ginzburg-Landau equation. Int J Mod Phys B 26 (06):1250035
14. Albert R, Barabasi A-L (2002) Statistical mechanics of complex networks. Rev Mod Phy 74:47–97
15. Freeman WJ, Kozma R, Bollobas B, Riordan O (2009) Chapter 7. Scale-free cortical planar network. In: Bollobas B, Kozma R, Miklos D (eds) Handbook of large-scale random networks. Series: Bolyai mathematical studies, vol 18. Springer, New York, pp. 277–324
16. Freeman WJ, Kozma R, Vitiello G (2012) Adaptation of the generalized Carnot cycle to describe thermodynamics of cerebral cortex. In: The 2012 International Joint Conference on Neural Networks (IJCNN). IEEE, pp 1–8

Commentators

Prof. Kazuyuki Aihara
Institute of Industrial Science
University of Tokyo
4-6-1, Komaba, Meguro-ku
Tokyo 153-8505, Japan
aihara@sat.t.u-tokyo.ac.jp

Dr. Bernard J. Baars, Ph.D., CEO
Society for Mind-Brain Sciences
The Neurosciences Institute
La Jolla, CA, USA
baarsbj@gmail.com
http://www.MBScience.org

Prof. Steven L. Bressler, Ph.D.
Center for Complex Systems and Brain Sciences
Department of Psychology, 777 Glades Road
Florida Atlantic University
Boca Raton, FL 33431, USA
bressler@fau.edu

Dr. Ray Brown, President
EEASI, 2100 Winrock, #64
Houston, TX 77057, USA
raybrown.easi@gmail.com

Prof. Péter Érdi, Henry R. Luce
Center for Complex Systems Studies
1200 Academy Street

Kalamazoo College
Kalamazoo, MI 49006, USA
Peter.Erdi@kzoo.edu

Prof. Morris W. Hirsch, Emeritus UCB
7926 Albe Road
Cross Plains, WI 53528-9350, USA
mwhirsch@chorus.net

Dr. Timothy Leleu
Institute of Industrial Science
University of Tokyo
4-6-1, Komaba, Meguro-ku
Tokyo 153-8505, Japan

Prof. Hans Liljenström
Agora for Biosystems, Director
Department of Energy and Technology
SLU, P.O. Box 7032
750 07 Uppsala, Sweden
Hans.Liljenstrom@slu.se

Prof. Dr. Frank W. Ohl
Leibniz Institute for Neurobiology
Department of Systems Physiology of Learning
Brenneckestr. 6
D-39118 Magdeburg, Germany
frank.ohl@ifn-magdeburg.de

Dr. Zoltan Somogyvari
Wigner Research Center for Physics
Hungarian Academy of Sciences
Konkolyi Thege M. ut 29-33
H-1121 Budapest, Hungary

Prof. Ichiro Tsuda
Research Institute for Electronic Science
Hokkaido University
Kita 12 Nishi 6, Kita-ku
Sapporo 060-0812, Japan
Tsuda@math.sci.hokudai.ac.jp

Prof. Giuseppe Vitiello, Ph.D.
Department of Physics “E.R. Caianiello”
University of Salerno
Baronissi (SA) 84100, Italy
vitiello@sa.infn.it

Prof. Paul J. Werbos
Department of Mathematical Sciences
University of Memphis
Memphis, TN 38152, USA
paul.werbos@gmail.com

Prof. James J. Wright
Department of Psychological Medicine
University of Auckland School of Medicine
Auckland, New Zealand
jj.w@xtra.co.nz

Contents

Part I
Review of Dynamical Brain Theories and Experiments

Chapter 1
Introduction—On the Languages of Brains

1.1 Brains Are Not Computers

The invention of digital computers over half a century ago fascinated scientists with enormous opportunities created by this new research tool, which potentially paralleled the capabilities of brains. Von Neumann has been one of the pioneers of this new digital computing era. While appreciating potential of computers, he warned about a mechanistic parallel between brains and computers. In his last work about the relationship between computers and brains, he pointed out that the operation of brains can not obey the potentially very high precision of algorithms postulated by Turing machines [1], and thus it is absolutely implausible that brains would use such algorithms in their operations. At higher levels of abstraction, in the last pages of his final work, Von Neumann contends that the language of the brain can not be mathematics as we know it [2]. He continues:

> It is only proper to realize that language is a largely historical accident. The basic human languages are traditionally transmitted to us in various forms, but their very multiplicity proves that there is nothing absolute and necessary about them. Just as languages like Greek and Sanskrit are historical facts and not absolute logical necessities, it is only reasonable to assume that logics and mathematics are similarly historical, accidental forms of expression. They may have essential variants, i.e. they may exist in other forms than the ones to which we are accustomed. Indeed, the nature of the central nervous system and of the message systems that it transmits, indicate positively that this is so. We have now accumulated sufficient evidence to see that whatever language the central nervous system is using, it is characterized by less logical and arithmetic depth than what we are normally used to. (Von Neumann, 1958)

If the language of the brain is not mathematics, if it is not a precise sequence of well-defined logical statements such as used in mathematics, then what is it? Von Neumann was unable to elaborate on this question due to his early tragic death. Half a century of research involving artificially intelligent computer designs could not give the answer either. This is partly due to the fact that Von Neumann's warning about principal limitations of the early designs of digital computers, called today 'Von Neumann

R. Kozma and W.J. Freeman, *Cognitive Phase Transitions in the Cerebral Cortex – Enhancing the Neuron Doctrine by Modeling Neural Fields*,
Studies in Systems, Decision and Control 39, DOI 10.1007/978-3-319-24406-8_1

computer architectures,' fell on deaf ears. Biological and human intelligence uses different ways of operations from the one implemented in symbol-manipulating digital computers. Nevertheless, researchers excited by seemingly unlimited power of digital computers embarked on projects of building increasingly complex computer systems to imitate and surpass human intelligence, without imitating natural intelligence. These projects gave impressive results, but notoriously fell short of producing systems approaching human intelligence.

The past half century produced crucial advances in brain research, in part due to advances in experimental techniques. An important challenge has been to reconcile the apparent contradiction between the absence of symbolic representations in brains as evidenced by neurodynamics and the symbolic nature of higher-level cognition and consciousness. In philosophy of artificial intelligence this is addressed as the symbol grounding problem.

The dynamical approach to cognition considers brain as a dynamic system moving along a complex non-convergent trajectory influenced by the subject's past and present experiences and anticipated future events. The trajectory may rest intermittently, for a fraction of a second, at a given spatio-temporal pattern. This pattern may have some meaning to the individual based on previous experiences. In this sense one may call this pattern a representation of the meaning of the given sensory influence in the context of the present internal state. However, the spatio-temporal pattern is unstable. A swift phase transition destroys it and moves the system along the trajectory. In other words, the quasi-stable spatio-temporal patterns can be considered as the words, and the phase transitions among patterns as the grammar of the brain code during the never ending cycles of cognitive processing.

The rest of this chapter describes the conceptual and philosophical challenges toward human brains and intelligence and their possible resolution.

1.2 Symbolic Approaches to Brains

The use of symbolic dynamics to study knowledge and cognition, which proved to be a rich source of powerful concepts dominating the field from the 60s through the 80s. The physical symbol system hypothesis illustrates key components of the symbolic approach [3–5]. According to this hypothesis, a physical symbol system has the necessary and sufficient means for intelligent action. In practical terms this means that the types of syntactic manipulation of symbols found in formal logic and formal linguistic systems typify this view of cognition. In this viewpoint, external events and perceptions are transformed into inner symbols to represent the state of the world. This inner symbolic code stores and represents all of the system's long-term knowledge. Actions take place through the logical manipulation of these symbols. This way, solutions are found for current problems presented by the environment. Problem solving takes the form of a search through a problem space of symbols, and the search is performed by logical manipulation of symbols through predefined operations (copying, conjoining, etc.). These solutions are implemented by forming plans and sending commands to the motor system to execute the plans to solve the problem. For an overview, see [6].

According to symbolic viewpoint, intelligence is typified by and resides at the level of deliberative thought. Modern examples of systems that fall within this paradigm include SOAR [7] and ACT-R [8]. The symbolic approach models certain aspects of cognition, and is capable of providing many examples of intelligent behavior. However, challenges to this viewpoint of cognition have appeared, both as practical criticisms of the performance of such systems and more philosophical challenges to the physical-symbol system hypothesis. On the practical side, symbolic models are notoriously inflexible and difficult to scale up from small and constrained environments to real world problems.

If symbolic systems are necessary and sufficient for intelligent behavior, why do we have such problems in producing the flexibility of behavior exhibited by biological organisms?

On the philosophical side, Dreyfus' situated intelligence approach is a prominent example of a criticism of symbolism. Dreyfus ascertains, following Heidegger's and Merleau-Ponty's traditions, that intelligence is defined in the context of the environment, therefore, a preset and fixed symbol system can not grasp the essence of intelligence [9, 10]. Pragmatic implementations of situated intelligence find their successful implementations in the field of embodied intelligence and robotics [11]. More recently, Dreyfus points out that …"Brooks' animates and all other versions of what some call Heideggerian AI have their own version of the frame problem, viz. that the program can not update relevance" [12]. He indicates that approaches based on dynamical systems theory can provide an account of how the brain of an active animal can directly pick up and update what counts as significant in its world. It is the hope of the authors of this essay that the described neurodynamical principles and mathematical models give a stimulation for the development of practically relevant autonomous devices in the spirit of Merleau-Ponty and Heidegger [13].

1.3 Connectionism

The connectionist view of cognition provides an alternative theory of the mind to the symbolic approach. Connectionist models emphasize parallel-distributed processing, while symbolic systems tend to process information in a serial fashion. Connectionist approaches represent adaptive and distributed structures, while symbols are static localized structures. Connectionist models offer many attractive features when compared with standard symbolic approaches. They have a level of biological plausibility absent in symbolic models that allows for easier visualization of how brains might process information. Parallel-distributed representations are robust, and flexible. They allow for pattern completion and generalization performance. They are capable of adaptive learning. In short, connectionist models provide a useful model of cognition, which are in many ways complementary to symbolic approaches.

Clark [14] categorizes modern connectionism into three generations, as listed below. We also add the fourth generation as the newest development in the field:

- First-generation connectionism: It began with the perceptron and the work of the cyberneticists in the 50s. It involves simple neural structures with limited capabilities. Their limitations draw criticism by representatives of the symbolist AI school in the 60s, which resulted in abandonment of connectionist principles by mainstream research establishment for decades. The resistance to connectivist ideas is understandable; it is in fact a repetition of millennia-old philosophical shift from nominalism to realism [15]. Connectionism has been revived in the mid '80s, thanks to the activities of the PDP research groups work (among others) on parallel distributed processing [16].
- Second-generation connectionism: It gained momentum since the '80s. It extends first-generation networks to deal effectively with complex dynamics of spatio-temporal events. It involves advanced recurrent neural network architectures and a range of advanced adaptation and learning algorithms. For an overview, see [17].
- Third-generation connectionism: It is typified by even more complex dynamic and time involving properties [18]. These models use biologically inspired modular architectures, along with various recurrent and hard-coded connections. Because of the increasing emphasis on dynamic and time properties, third-generation connectionism has also been called dynamic connectionism. Third generation connectionist models include DARWIN [19], and the Distributed Adaptive Control (DAC) models [15, 20].
- Fourth generation connectionism: The newest development of neural modeling, representing an additional step going beyond Clark's original categorization schema. This approach involves nonconvergent/chaotic sequences of spatio-temporal oscillations [21, 22]. It is based on advances in EEG analysis, which gave spatiotemporal amplitude modulation (AM) patterns of unprecedented clarity. The K (Katchalsky) models are prominent examples of this category, which are rooted in intuitive ideas from the 70s [23] and gained prominence since the turn of the century.

Our focus is new developments in fourth generation connectionism. A key to these models is the mesoscopic-intermediate-range paradigm [24]. Accordingly, intelligence in brains is rooted in the delicate balance between local fragmentation of individual components at the cellular level, and overall dominance of a unified global component at the brain/hemisphere level. This balance is manifested through metastable dynamic brain states undergoing frequent state transitions. Phase transitions are crucial components of the new generation of connectionist models as they provide vehicles to produce the seamless sequence of spatio-temporal oscillation patterns punctuated by cortical phase transitions. This is the main focus of this work.

Traditional approaches to knowledge extraction and ontology generation in machine learning and trained systems have focused on static systems, extracting grammatical rules from dynamical systems or creating ontologies primarily from text or numerical data databases [18]. These approaches are inherently limited by nature of extracted knowledge representations, which are static structures with no dynamics. Dynamic approach to intelligence creates the opportunity of integrating bottom–up and top–down methods of intelligence. It goes beyond bottom–up connectionist approaches by extracting symbolic knowledge from sub-symbolic structures manifested in the form of spatio-temporal fluctuations of brain activity.

1.4 Brains as Transient Dynamical Systems

There is a rich literature aiming at bridging the difference in spatial and temporal scales between microscopic properties of single neurons and macroscopic properties of large populations of neurons, including coarse graining, mean-field approaches, circular causality [23, 25–27]. However, there is a vast amount of theoretical issues to be resolved before understanding the operation of brain as a unified organ. Interactive populations of neurons operating far from equilibrium create spatiotemporal patterns of activity that sustain intelligent behaviors [28–30]. These patterns are dissipative structures, because they require prodigious quantities of metabolic energy, also called *dark energy* [31].

Measurements of electrical potential fields in brains document the presence of spatial textures of activity patterns; the experimental method can be either invasive using electrocorticogram (ECoG) electrodes on the cortical surface [32], or noninvasive electroencephalogram (EEG) using scalp electrodes [33]. Each pattern has a narrow-band carrier frequency of oscillation in the beta-gamma range. The patterns are formed by the modulation of the amplitude (AM) and phase (PM) of the carrier wave. The sizes of AM and PM patterns vary from a few mm in diameter (ECoG) [34] to the entire scalp (EEG) [33]. The temporal spectra are power-law [35]. The distributions of the durations of the patterns are power-law [36]. These findings imply that the AM-PM patterns are scale-free in space and time across the beta and gamma ranges [28, 37–39]. The AM and PM patterns are represented as n-dimensional vectors in the feature space, where n is the number of pixels in the patterns. The vector from multichannel ECoG recordings serves as a vectorial state variable in modeling. These basic concepts provided the foundation for chaotic brain dynamics, following the pioneering work by [40].

Following Prigogine and Haken theory on emergent behavior in open thermodynamic systems [41, 42], brains have been modeled as a dissipative thermodynamic system that by homeostasis holds itself near a critical level of activity that is a non-equilibrium metastable state. Principles of self-organization and metastability have been introduced to describe brain dynamics [43, 44]. Recently, the concept of self-organized criticality (SOC) has been employed in neuroscience to provide a model framework for the observations [45, 46]. There is empirical evidence of cortex conforming to the self-stabilized near critical state during the existence of quasi-stable states [47–49]. Brains exhibit transient switches between the meta-stable states. SOC, however, cannot produce the sequence of transient patterns observed in cognitive experiments [28, 50, 51]. Bonachela describes SOC as *pseudo-critical* [52] and suggests to complement self-organization with more elaborated, adaptive approaches.

Here we propose to use tools of random graph theory (RGT) [53–56] and percolation dynamics [57] to model critical behavior in brain networks [58]. When modeling the cortical sheet, we consider graphs in the geometric space, in our case over a two dimensional lattice. The corresponding mathematical objects are called cellular automata, which can be either deterministic or probabilistic (PCA). We build on the wealth of research accumulated in the past decades on the structure and function of brain networks [59], including hubs and small worlds [60–62], identification of percolation transitions in living neural networks [63–65], the presence and absence of scale-free properties [66, 67], causal links [68, 69], and *Rich Club* networks [70–72]. We explore how the identified brain network properties contribute to critical spatio-temporal dynamics with discontinuous transitions during cognitive processing.

1.5 Random Graph Theory (RGT) for Brain Models

Dynamics of neural populations are widely modeled by nonlinear ordinary differential equations (ODEs) with distributed parameters and by partial differential equations (PDEs). The nonlinear equations may be solved by piecewise linear approximations [23, 25, 73], or by numerical integration. In the latter case noise plays a key role in stabilizing chaotic trajectories across the extremely fragmented attractor basins [24, 74]. Classical tools of calculus and integro-differential equations have their limitations. Namely, obtaining stable solutions for large systems of nonlinear ODEs describing ECoG activity is often an extremely difficult and time-consuming task on both digital and analog platforms [75, 76].

Von Neumann wrote that brains do in a "few short steps" what computers do with exquisite numerical precision over many logical steps, because "brains lack the arithmetic and logical depth that characterize computers." He added: "Whatever the system is, it cannot fail to differ considerably from what we consciously and explicitly consider as mathematics" ([2], pp. 80–81). Von Neumann's observation about the deep divide between the operation of brains and the existing tools of mathematical analysis underlines the need for new approaches to define and model these *few short steps* in brains. In addition, he indicated the potential direction of new math through his work on cellular automata and finite mathematics.

In the past 60 years, cellular automata have developed into powerful mathematical tools capable of modeling complex spatio-temporal dynamics and emergent structures. Due to his untimely death from osteosarcoma von Neumann was unable to participate in the development of this new branch of mathematics. The seminal work by Alan Turing moved beyond symbolic theory of computation, including his own work on Turing machines for artificial intelligence [1]. Following Turing's work on morphogenesis, pattern formation has been described by diffusion-coupled chemical reactions with very rich spatio-temporal dynamics, such as the Belousov-Zhabotinsky and Liesegang patterns [77].

Systems theory has been introduced to describe emergence in living organisms [78] and in biochemical processes [73]. Ilya Prigogine modeled the emergence of structure in open chemical systems operating far from thermodynamic equilibrium [41]. He called the resulting patterns dissipative structures, because they emerged by the interactions of particles feeding on energy with local reduction in entropy. Hermann Haken developed the field of synergetics [42], which is the study of pattern formation in lasers by particles, whose interactions create order parameters by which the particles govern themselves in circular causality. Various approaches have been developed to extend these results to neural systems, including dynamical systems theory of brains and behaviors [40, 79].

Random graphs provide a mathematical framework to describe phase transitions and critical phenomena in large-scale brain networks and they have distinct advantages as compared to ODEs and PDEs when modeling cortical spatio-temporal transitions. One advantage is obvious: the systems of nonlinear differential equations are typically solved in discretized form in space and time, with the exception of rare instances of specific problems, when analytic solutions can be derived. Moreover, differential equations require some degree of smoothness in the modeled phenomena, which may not be suitable to describe the observed sudden changes in neurodynamics.

Two contradictory aims are to be reconciled when developing mathematical models of complex systems, such as the cortex, using graph theory. The developed model should produce random graphs whose structure resembles that of the neocortex as much as possible, on the one hand, but it should not be so complicated that it is not susceptible to mathematical analysis, on the other hand. The hope is that the mathematical theory of random graphs will eventually be advanced enough that both requirements can be satisfied: a sophisticated model can be constructed that incorporates most of the requirements, and this very complicated model can be analyzed precisely, meaning that one can obtain mathematical descriptions both of its structure, and of the associated dynamics. RGT provides a solid foundation to describe the evolution of networks of continuously increasing size that can approach and exceed dynamic boundaries, thereby bridging the divide between the microscopic and macroscopic domains by summing rather than approximating.

1.6 Neuropercolation Modeling Paradigm

Neuropercolation generalizes probabilistic cellular automata and percolation theory in random graphs that are structured in accordance with cortical architectures, and it has the potential of effectively describe the sudden changes and discontinuities observed in cortical processes [13, 80]. Neuropercolation develops a theory of neurodynamics based on self-organization in dissipative structures [80–84]. Neuropercolation models collective properties of brain networks near critical states, when the behavior of the system changes abruptly with the variation of some parameter. We use the hierarchy of interactive populations in brain networks as described by Freeman K models [23] and replace ODEs with PCAs. Hypotheses are outlined about the rapid propagation of phase gradients in the cortex during singular phase transitions, and the requisite structure of the cortical tissue evolving explosively during the ontogenetic development of infants.

Neuropercolation approach links pattern-based spatio-temporal encoding and generalized, non-local percolation in random graphs. The corresponding theory is applied for modeling the dynamics of sensor networks during sensing and decision making. The connectivity between components of a sensor network greatly influences the system's ability to respond to changes. Percolation theory shows that the relationship between the individual connectivity of the components and the systems collective response to changes is not a smooth function. Thorough analysis can identify critical conditions when the system overall behavior changes drastically in response to relatively small variations in local behaviors.

Recent observations concerning the special role of extreme events across disciplines, including stock market crashes, solar flares, earthquakes, the growth of megalopolises, and other natural phenomena [46] provide support to a new class of models, coined as *Dragon Kings (DK)*. According to DK phenomenology, certain extreme events are outliers rather than extensions of the tail of scale-free probability distributions at very low frequencies [85–87]. The neuropercolation model of critical phase transitions in brains as large-scale random networks is in line with major premises of DK phenomenology [51, 88]. In this monograph, we introduce neuropercolation as a new mathematical tool to supplement differential equations, especially in modeling experimentally-observed rapid changes in the space-time behavior of brains.

References

1. Turing AM (1954) The chemical basis of morphogenesis. Philos Trans R Soc Lond B 237:37–94
2. Von Neumann J (1958) The computer and the brain. Yale UP, New Haven
3. Newell A, Simon HA (1972) Human problem solving. Prentice-Hall, Englewood Cliffs
4. Newell A (1980) Physical symbol systems. Cogn Sci 4:135–183
5. Newell A (1990) Unified theories of cognition. Harvard University Press, Cambridge, MA
6. Harter D, Kozma R (2006) Aperiodic dynamics and the self-organization of cognitive maps in autonomous agents. Int J Intell Syst 21(9):955–972
7. Laird JE, Newell A, Rosenbloom PS (1987) SOAR: an architecture for general intelligence. Artif Intell 33:1–64
8. Anderson JA, Silverstein JW, Ritz SR, Jones RS (1977) Distinctive features, categorical perception, and probability learning: some applications of a neural model. Psychol Rev 84:413–451
9. Merleau-Ponty M (1945) Phnomnologie de la perception. Gallimard, Paris
10. Dreyfus HL (1992) What computers still can't do—a critique of artificial reason. MIT Press, Cambridge
11. Mataric MJ, Brooks RA (1999) Learning a distributed map representation based on navigation behaviors. In: Brooks RA (ed) Cambrian intelligence. MIT Press, Cambridge, pp 37–58

12. Dreyfus HL (2009) How representational cognitivism failed and is being replaced by body/world coupling. After cognitivism: a reassessment of cognitive science and philosophy. Springer, New York, pp 39–73
13. Kozma R, Freeman WJ (2009) The KIV model of intentional dynamics and decision making. Neural Netw 22(3):277–285
14. Clark A (2001) Mindware: an introduction to the philosophy of cognitive science. Oxford University Press, Oxford
15. Pfeifer R, Scheier C (1999) Understanding intelligence. MIT Press, Cambridge
16. Rumelhart DE, McClelland JL (1986) Parallel distributed processing: explorations in the microstructure of cognition. MIT Press, Cambridge
17. Haykin S (1998) Neural networks—a comprehensive foundation. Prentice Hall, New Jersey
18. Towell GG, Shavlik JW (1994) Knowledge-based artificial neural networks. Artif Intell 70:119–165
19. Edelman GM, Tononi G (2000) A universe of consciousness: how matter becomes imagination. Basic Books, New York
20. Vershure PM, Althaus P (2003) A real-world rational agent: unifying old and new AI. Cogn Sci 27(4):561–590
21. Crutchfield J (1990) Computation at the onset of chaos. In: Zurek W (ed) Entropy, complexity, and the physics of information. Addison-Wesley, Reading, pp 223–269
22. Kaneko K, Tsuda I (2001) Complex systems: chaos and beyond. A constructive approach with applications in life sciences. Springer, New York
23. Freeman WJ (1975/2004) Mass action in the nervous system. Academic, New York. Electronic version 2004. http://sulcus.berkeley.edu/MANSWWW/MANSWWW.html
24. Kozma R, Freeman WJ (2001) Chaotic resonance: methods and applications for robust classification of noisy and variable patterns. Int J Bifurc Chaos 10:2307–2322
25. Katchalsky A, Rowland V, Huberman B (1974) Dynamic patterns of brain cell assemblies. Neurosci Res Prog Bull 12:1–187
26. Fingelkurts AA, Fingelkurts AA (2004) Making complexity simpler: multivariability and metastability in the brain. Int J Neurosci 114:843–862
27. Freeman WJ, Quian Quiroga R (2013) Imaging brain function with EEG: advanced temporal and spatial analysis of electroencephalographic and electrocorticographic signals. Springer, New York
28. Freeman WJ (2008) A pseudo-equilibrium thermodynamic model of information processing in nonlinear brain dynamics. Neural Netw 21:257–265
29. Freeman WJ, Livi R, Obinata M, Vitiello G (2012) Cortical phase transitions, nonequilibrium thermodynamics and time-dependent Ginzburg-Landau equation. Int J Mod Phys B 26(06):1250035
30. Freeman WJ, Kozma R, Vitiello G (2012) Adaptation of the generalized Carnot cycle to describe thermodynamics of cerebral cortex. In: The 2012 international joint conference on neural networks (IJCNN). IEEE, pp 1–8
31. Raichle ME (2006) The brain's dark energy. Science 314:1249–1250
32. Barrie JM, Freeman WJ, Lenhart M (1996) Modulation by discriminative training of spatial patterns of gamma EEG amplitude and phase in neocortex of rabbits. J Neurophysiol 76:520–539
33. Ruiz Y, Pockett S, Freeman WJ, Gonzales E, Guang L (2010) A method to study global spatial patterns related to sensory perception in scalp EEG. J Neurosci Methods 191:110–118
34. Freeman WJ, Baird B (1987) Relation of olfactory EEG to behavior: spatial analysis. Behav Neurosci 101(3):393
35. Freeman WJ, Zhai J (2009) Simulated power spectral density (PSD) of background electrocorticogram (ECoG). Cogn Neurodyn 3(1):97–103
36. Freeman WJ (2004) Origin, structure, and role of background EEG activity. Part 2. Anal Phase Clin Neurophysiol 115:2089–2107
37. Stam CJ, de Bruin A (2004) Scale-free dynamics of global functional connectivity in the human brain. Hum Brain Mapp 22:97–109

38. Freeman WJ, Ahlfors SM, Menon V (2009) Combining EEG, MEG and fMRI signals to characterize mesoscopic patterns of brain activity related to cognition. Special Issue (Lorig TS ed). Int J Psychophysiol 73(1):43–52
39. Kello CT, Brown GD, Ferrer-i-Cancho R, Holden JG, Linkenkaer-Hansen K, Rhodes T, Van Orden GC (2010) Scaling laws in cognitive sciences. Trends Cogn Sci 14(5):223–232
40. Skarda CA, Freeman WJ (1987) How brains make chaos in order to make sense of the world. Behav Brain Sci 10:161–195
41. Prigogine I (1980) From being to becoming: time and complexity in the physical sciences. WH Freeman, San Francisco
42. Haken H (1983) Synergetics: an introduction. Springer, Berlin
43. Kelso JAS (1995) Dynamic patterns: the self organization of brain and behavior. MIT Press, Cambridge
44. Haken H (2002) Brain dynamics: synchronization and activity patterns in pulse-coupled neutral nets with delays and noise. Springer, New York
45. Beggs JM, Plenz D (2003) Neuronal avalanches in neocortical circuits. J Neurosci 23(35):11167–11177
46. De Arcangelis L (2012) Are dragon-king neuronal avalanches dungeons for self-organized brain activity? Eur Phys J Spec Top 205(1):243–257
47. Bak P (1996) How nature works the science of self-organized criticality. Springer, New York
48. Beggs JM (2008) The criticality hypothesis: how local cortical networks might optimize information processing. Philos Trans R Soc A: Math, Phys Eng Sci 366(1864):329–343
49. Petermann T, Thiagarajan TC, Lebedev MA, Nicolelis MA, Chialvo DR, Plenz D (2009) Spontaneous cortical activity in awake monkeys composed of neuronal avalanches. Proc Natl Acad Sci 106(37):15921–15926
50. Rabinovich MI, Friston KJ, Varona P (eds) (2012) Principles of brain dynamics. MIT Press, Cambridge
51. Kozma R, Puljic M (2015) Random graph theory and neuropercolation for modeling brain oscillations at criticality. Curr Opin Neurobiol 31:181–188
52. Bonachela JA, de Franciscis S, Torres JJ, Munoz MA (2010) Self-organization without conservation: are neuronal avalanches generically critical? J Stat Mech: Theor Exp 2010(02):P02015
53. Erdos P, Renyi A (1959) On random graphs. Publ Math Debr 6:290–297
54. Erdos P, Renyi A (1960) On the evolution of random graphs. Publ Math Inst Hung Acad Sci 5:17–61
55. Bollobas B (1985/2001) Random graphs, 2nd edn., Cambridge studies in advanced mathematics. Cambridge UP, Cambridge
56. Kauffman S (1993) The origins of order–self-organization and selection in evolution. Oxford UP, Oxford
57. Bollobas B, Riordan O (2006) Percolation. Cambridge UP, Cambridge
58. Bollobas B, Kozma R, Miklos D (eds) (2009) Handbook of large-scale random networks., Bolyai society mathematical studies. Springer, New York
59. Fellemin DJ, Van Essen DC (1991) Distributed hierarchical processing in the primate cerebral cortex. Cereb Cortex 1:1–47
60. Breakspear M (2004) Dynamic connectivity in neural systems: theoretical and empirical considerations. Neuroinformatics 2(2):205–225
61. Bassett DS, Meyer-Lindenberg A, Achard S, Duke T, Bullmore E (2006) Adaptive reconfiguration of fractal small-world human brain functional networks. PNAS 103(51):19518–19523
62. Sporns O, Honey CJ (2006) Small worlds inside big brains. Proc Natl Acad Sci 103(51):19219–19220
63. Breskin I, Soriano J, Moses E, Tlusty T (2006) Percolation in living neural networks. Phys Rev Lett 97(18):188102
64. Tlusty T, Eckmann JP (2009) Remarks on bootstrap percolation in metric networks. J Phys A: Math Theor 42(20):205004
65. Eckmann JP, Moses E, Stetter O, Tlusty T, Zbinden C (2010) Leaders of neuronal cultures in a quorum percolation model. Front Comput Neurosci 4(132). doi:10.3389/fncom.2010.00132

66. Hagmann P, Cammoun L, Gigandet et al (2008) Mapping the structural core of human cerebral cortex. PLOS Biol 6(7):e159, 1–14
67. Bullmore E, Sporns O (2009) Complex brain networks: graph theoretical analysis of structural and functional systems. Nat Rev Neurosci 10:1–13
68. Bressler S, Menon V (2010) Large-scale brain networks in cognition:emerging methods and principles. Trends Cogn Sci 14:277–290
69. Hu S, Wang H, Zhang J, Kong W, Cao Y, Kozma R (2015) Comparison Analysis: granger causality and new causality and their applications to motor imagery. IEEE Trans Neural Netw Learn Syst (in press)
70. Zamora-Lopez G (2009) Linking structure and function of complex cortical networks. Ph.D. thesis, University of Potsdam, Potsdam
71. Zamora-Lopez G, Zhou C, Kurths J (2011) Exploring brain function from anatomical connectivity. Front Neurosci 5:83
72. Van den Heuvel MP, Sporns O (2011) Rich-club organization of the human connectome. J Neurosci 31(44):15775–15786
73. Katchalsky Katzir A (1971) Biological flow structures and their relation to chemodiffusional coupling. Neurosci Res Prog Bull 9:397–413
74. Kozma R (2003) On the constructive role of noise in stabilizing itinerant trajectories on chaotic dynamical systems. Chaos 11(3):1078–1090
75. Principe JC, Tavares VG, Harris JG, Freeman WJ (2001) Design and implementation of a biologically realistic olfactory cortex in analog VLSI. Proc IEEE 89:1030–1051
76. Srinivasa N, Cruz-Albrecht JM (2012) Neuromorphic adaptive plastic scalable electronics: analog learning systems. IEEE Pulse 3(1):51–56
77. Zhabotinsky AM, Zaikin AN (1973) Autowave processes in a distributed chemical system. J Theor Biol 40:45–61
78. Von Bertalanffy L (1968) General system theory: foundations, development, application. George Braziller Press, New York
79. Freeman WJ (1991) The physiology of perception. Sci Am 264(2):78–85
80. Kozma R (2007) Neuropercolation. Scholarpedia 2(8):1360
81. Kozma R, Puljic M, Balister P, Bollobas B, Freeman WJ (2005) Phase transitions in the neuropercolation model of neural populations with mixed local and non-local interactions. Biol Cybern 92:367–379
82. Kozma R, Puljic M, Freeman WJ (2012) Thermodynamic model of criticality in the cortex based on EEG/ECoG data. arXiv preprint arXiv:1206.1108
83. Kozma R, Puljic M (2013) Learning effects in neural oscillators. Cogn Comput 5(2):164–169
84. Kozma R, Puljic M (2013) Hierarchical random cellular neural networks for system-level brain-like signal processing. Neural Netw 45:101–110
85. Johansen A, Sornette D (2010) Shocks, crashes and bubbles in financial markets. Bruss Econ Rev (Cahiers economiques de Bruxelles) 53(2):201–253
86. Sornette D, Quillon G (2012) Dragon-kings: mechanisms, statistical methods and empirical evidence. Eur Phys J Spec Top 205(1):1–26
87. Pisarenko VF, Sornette D (2012) Robust statistical tests of Dragon-Kings beyond power law distributions. Eur Phys J Spec Top 205(1):95–115
88. Erdi P, Kozma R, Puljic M, Szente J (2013) Neuropercolation and related models of criticalities. In: Contents XXIX-th european meeting of statisticians, Hungary, p 106

Chapter 2
Experimental Investigation of High-Resolution Spatio-Temporal Patterns

2.1 Method

In the past decades, a wide range of experiments have been conducted using invasive and non-invasive techniques to analyze neural correlates of cognitive behaviors. Here we introduce examples of high-resolution intracranial electrocorticogram (ECoG) experiments with rabbits and human participants, as well as noninvasive scalp electroencephalogram (EEG) experiments. Results of these experiments will be used to illustrate the neurodynamic principles of cognition discussed in this work.

2.1.1 Experiments with Rabbits

In the experiments with New Zealand rabbits, the animals have been chronically implanted epidurally with square arrays of 8×8 electrodes. The electrodes were 0.25 mm in diameter, they were made of stainless-steel, epoxy-insulated wires with connectors before surgery [1]. 8×8 arrays were positioned over the auditory cortex, visual cortex, olfactory bulb, and prepyriform cortex, respectively; see Fig. 2.1. Structural ground and reference screws were inserted into the skull for monopolar recordings after recovery from surgery. All procedures were conducted according to protocol approved by the University of California at Berkeley Animal Use and Care Committee with veterinary supervision by the Office of Laboratory Animal Care. During the experiments, each rabbit was placed in a restraining apparatus to reduce EEG artifacts due to excess body movement. The rabbits were placed in an electrically shielded chamber. The signals from the arrays were recorded using amplifiers with band-pass analog filters. The EEG was edited for movement artifacts and for stimulus- specific activity. The rabbits were trained in a classical aversive paradigm to discriminate a reinforced conditioned visual stimulus (CS+) from another not reinforced one (CS-). Each stimulus was delivered for 20 trials, randomly interspersed,

R. Kozma and W.J. Freeman, *Cognitive Phase Transitions in the Cerebral Cortex – Enhancing the Neuron Doctrine by Modeling Neural Fields*,
Studies in Systems, Decision and Control 39, DOI 10.1007/978-3-319-24406-8_2

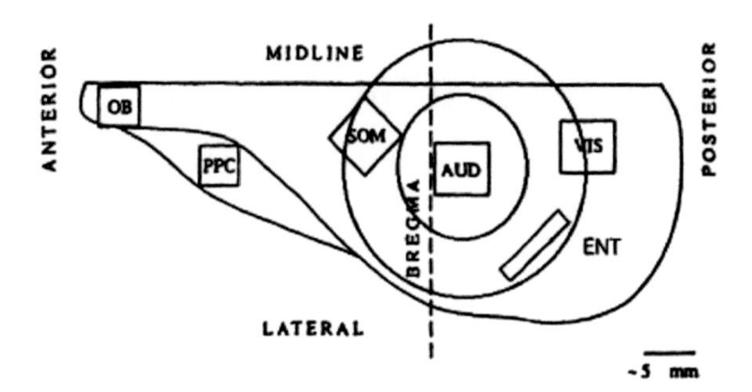

Fig. 2.1 Illustration of the location of the implanted array locations in rabbit multi-cortical experiments; OB olfactory bulb, PPC—prepyriform cortex, SOM—somatosensory cortex, AUD—auditory cortex, VIS—visual cortex. Each array consists of 4 4 electrodes; from [1]

for a total of 40 trials per experiment. The 6 s recording period was divided into 3 s pre-stimulus and 3 s post-stimulus EEG recording. These experiments were conducted in the 90s, and they provided massive amount of ECoG data with high spatial and temporal resolution, which have been thoroughly analyzed in the past decades in search for neural correlates of cognition, such as learning and classification.

The ECoG recorded from 8 × 8 high-density arrays reveals high correlation among the 64 signals. Commonly the correlation is attributed to activity at the reference electrode in referential recording and to volume conduction. These factors do not account for the high correlation, as shown by the 2-fold differences in amplitude and phase of the shared carrier waveform often seen between signals from adjacent electrodes, and as proved by calculation of the dendritic point spread function of the cortical generators [2].

2.1.2 Human ECoG Experiments

Very few experimental results are available with high-density ECoG arrays in human subjects. Attempts were made to replicate the classification results in humans through subdural recordings of gamma activity from the exposed somatosensory cortex under local anesthesia in neurosurgical patients, who were undergoing treatment for intractable epilepsy. However, the large spacing (1 cm) of the electrodes lead to unsatisfactory results [3]. This value was close to the upper limit of the estimated size of coherent domains in animal studies [4].

Here we introduce measurements with 1× 64 array was designed with spacing of 0.5 mm between stainless steel electrodes 0.1 mm in diameter, giving a Nyquist frequency of 1.0 c/mm. The array was 32 mm in length, giving a lower spatial spectral limit of 0.031 c/mm, so it could be applied to a single gyrus, typically in humans 1 cm in width and several cm in length, without crossing intervening sulci.

ECoG were recorded from the 64 electrodes in a linear array 3.2 cm long, which was placed on the exposed superior temporal gyrus or motor cortex of volunteers undergoing diagnostic surgery. The experiments have been conducted at the Neurophysiology Laboratory, Harborview Medical Center, Seattle WA [5]. Visual

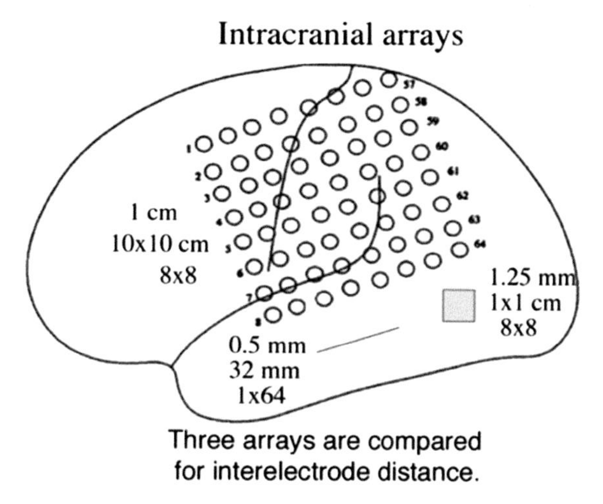

Fig. 2.2 Illustration of several possible intracranial experimental array designs; a large square array of 8 × 8 electrodes with 1 cm spacing; a high-density 8 × 8 array with 1.25 mm spacing; linear array of 64 electrodes with 0.5 mm spacing

displays of multiple traces revealed broad spectrum oscillations in episodic bursts having a common aperiodic wave form with recurring patterns of spatial amplitude modulation (AM patterns) on selected portions of the array. The one dimensional spatial spectrum of the human ECoG was calculated at successive time samples and averaged over periods up to 20 s.

In Fig. 2.2 several possible intracranial experimental array designs are compared. The 8 × 8 square array measuring 8 × 8 cm has electrodes spreading over the frontal, temporal and parietal lobes at intervals of 1 cm [3]. High spatial resolution linear and square arrays are indicated as well [5, 6].

2.1.3 Scalp EEG Design Considerations

Experiments with noninvasive scalp EEG arrays are highly desirable and there has been rapid progress in the field since early pioneering work [7–9]. Scalp EEG is difficult to measure due to contamination with electromyographic (EMG) potentials having similar temporal spectral properties. Substantial improvements in signal identification techniques applied to EEG records have provided strong evidence for the correlation of brief bursts of gamma EEG during cognitive processing in perception and decision making by human volunteers [10–14]. Its significance is thought to lie in manifesting neural operations by which "binding" of neural activity occurs between networks of neurons separated by relatively great distances in the forebrain, particularly the cerebral cortex of both hemispheres [8, 15–18].

There are still relatively few experimental designs with high-resolution spatial arrays for monitoring cognitive functions in human subjects [6]. Thorough simulation of electrical fields at the scalp and comparison between EEG, MEG and structural MRI results, the optimal arrangement of scalp EEG electrodes can be estimated.

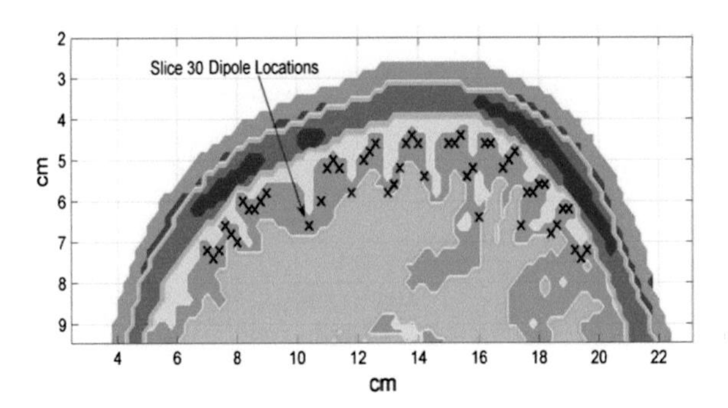

Fig. 2.3 Display of a segmented image slice with the identified major tissues: scalp, hard and soft skull bone, cerebrospinal fluid (CSF), cortical *gray* matter, and *white* matter. The cortical contour formed by the boundary of CSF and the gray matter was used for dipole locations to model the electrical activity. The same contour was used for computing the PSD of the cortical contours. Only the *top* portion of the contour is shown above 9 cm, which was used for modeling [19]

Here we introduce the approach by [19, 20] where a finite element method (FEM) model of the head was used based on a well-established MR imaging approach. Sagittal MRI slices of an adult male subject with 3.2 mm thickness were collected with a 1.5 Tesla GE Signa scanner [20–22]. The MR images were segmented using a semiautomatic tissue classification program. The identified tissues were: scalp, fat, muscle, hard skull bone, soft skull bone, gray matter, white matter, eyes, spinal cord and cerebellum, cerebrospinal fluid (CSF) and soft tissue. One of the segmented slices, marked as slice number 30 is shown in Fig. 2.3. This is the 30th slice starting from the left side of the subject. It is 1.7 cm to the right from the midline of the brain. The x coordinate increases from anterior (front) to the posterior (back) of the subject, the y coordinate increases from superior (top of the head) to the inferior (bottom) and the z coordinate increases from left to the right side of the subject.

The MR results have been used to simulate EEG signals. Tissue resistivity values used in the model were compiled from published values [23–25] following the methodology described in [20, 22]. Using a uniform finite element solver [26], the scalp potentials were computed due to dipolar sources confined to the gray matter in the upper portion of the cerebral hemispheres. Spatial and temporal power spectral densities (PSD) at the scalp have been calculated based on the simulated EEG and indicate an optimal spatial electrode distance of about 3–5 mm [19].

2.2 Temporal Patterns: The Carrier Wave

Mutually excitatory neural populations are the backbone of cortical dynamics, because the positive feedback among excitatory neurons provides the cortical background activity. The connection density increases steadily during embryological development until each neuron gets back as many pulses on average as it transmits

[27]. The feedback gain equals unity and the cortical activity is self-sustaining. Neurons must be continually active to survive and flourish, thus the transition to self-sustaining activity is of great value. As the level of connectivity continues to rise, the refractory periods prevent the output from increasing further and produce a steady state. The time course of the signal fed back to each neuron from the polysynaptic matrix of connections can be approximated by a random diffusion process [16] and the activity in every local area is randomized. The background spontaneous noise enables brains to maintain an active state of readiness to search the environment and initiate action in any desired direction. The noise provides the matrix from which the cortex creates non-random, structured patterns in response to sensory input by organizing and condensing the random background. In brief, the achievement of sustained spontaneous background activity by mutual excitation is the mother of all cortical state transitions by which brains execute intentional actions.

The power spectral density distribution (PSD) of cortical processes is linear in the coordinates of log power vs. log frequency, $1/f^{\alpha}$, conforming to a power-law distribution. Were it not for the refractory periods, the slope of the line would have an exponent, $\alpha \approx 2$ (brown noise) [28]. However, the refractory periods act as a low-pass filter that increases the exponent, $2 < \alpha < 4$ (black noise) by blocking high-frequency firing [29]. Under perturbation by sensory and centrifugal input to cortex the activity is driven above or below the set point, but it tends to return to the set point upon cessation of the input. The level of the steady-state noise is regulated by brain stem nuclei whose neuromodulators (acetylcholine, norepinephrine, histamine) determine the level of behavioral arousal and the amplitude of the ECoG [16], Chap. 7.

Examples of PSDs for human and rabbit EEG are displayed in Fig. 2.4. The temporal PSDT displayed a range with nearly linear decrease in log power with increasing log frequency ($1/f^{\alpha}$, where α 2 on average). The band pass temporal filter required for the Hilbert transform was placed in the beta-gamma range (12–55 Hz).

In a population of inhibitory neurons, the units interact by mutual inhibition, which is a form of positive feedback (double negative sign). The classic example is the layer of interneurons in the eye of the horseshoe crab Limulus, which is responsible for spatial contrast and the formation of Mach bands [31]. The frequency of the oscillation is mainly determined by the open loop passive membrane time constants of the component neurons, typically 5 ms. Each cycle requires four steps around the loop, giving a median value of 50 Hz, but with modulation of the frequency by the two forms of positive feedback to give values through the upper beta range and all of the gamma range (20–80 Hz).

A common misconception is to identify positive feedback with instability and negative feedback with stability. Actually both types of interaction are stable at weak levels of interaction, and they both have thresholds for instability above some critical value of interaction strength, when it is expressed as a feedback gain. The difference is that positive feedback gives sustained, monotonic increases in activity, whereas negative feedback gives sustained oscillations, such as the bursts of beta and gamma commonly seen in the ECoG and EEG. Another common misconception is

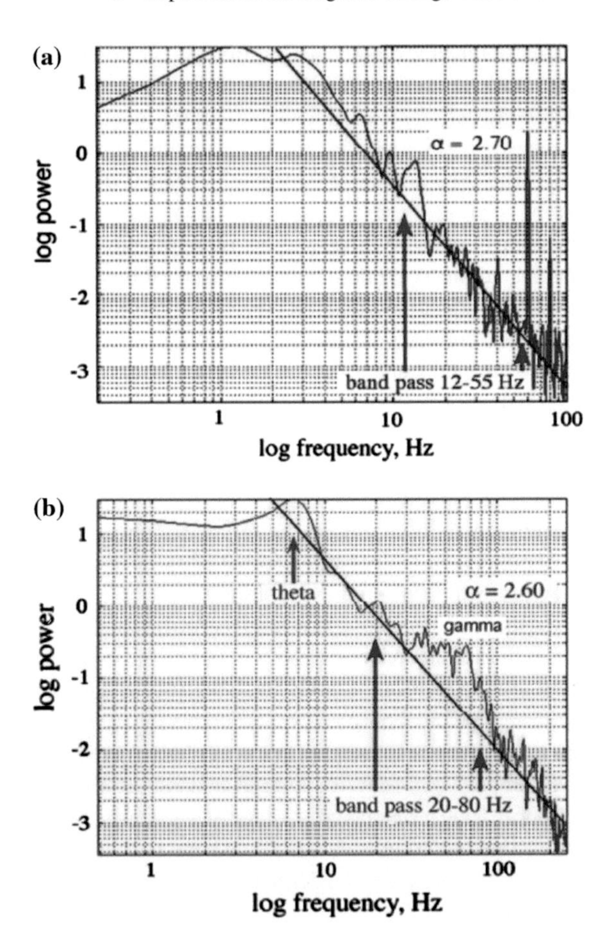

Fig. 2.4 Illustration of the average temporal power spectral density (PSDt) inhuman and animal experiments. **a** PSDt of human subject at rest; it was calculated over an epoch of 1000 digitizing steps (5 s) for each of the 64 channels and averaged across channels, N = 64. ([2]. **b** Average PSDt for active rabbit; the spectrum has a $1/f^{\alpha}$ shape with excess power in the theta (3–7 Hz) and beta-gamma (20–80 Hz) ranges. [30]

that the sustained interaction from positive feedback is only a steady bias with zero frequency. It is that, but it is more, because its spectrum shows that it contains all frequencies in the ECoG/EEG range.

2.3 Spatial Patterns of Amplitude Modulation (AM) and Phase Modulation (PM)

Here we present results with EEG and ECoG measurements. High-resolution recordings from multiple electrodes in square arrays, fixed on cortical surfaces provide insight to the basic structure of ECoG [6]. The ECoG reveals a form of what is called black noise [28] which is a random variable having intermittent, non-Gaussian extreme values [32] and from which evanescent, non-random structures emerge in the temporal and spatial domains. These structures are the correlates of the perceptual and cognitive events that we seek to understand. We apply synchronization measures [33] to study collective neural dynamics is space and time.

The decomposition of ECoG and EEG signals into amplitude and phase functions is by the Fourier transform giving high resolution of frequency in stationary signals, and by the Hilbert transform giving high temporal resolution of signals undergoing phase and frequency modulation; see Appendix for details. The amplitude values have behavioral correlates; the phase values do not, but they reveal the manner and mechanism by which cortical dynamics makes patterns.

A spatial ECoG pattern was defined as a set of digitized amplitudes at a given point in time. What made this meaningful was the finding that metastable frames emerged in the ECoG, while the signals had the same waveform at a nearly constant, shared carrier frequency in the alpha-theta band. The amplitudes were measured by calculating the root mean square amplitude of the band-pass filtered signal [1], or by calculating with Hilbert transform the maximal mean analytic amplitude of the filtered signals [34]. Owing to the shared carrier frequency, each AM pattern can be displayed in a contour plot showing peaks and valleys. AM patterns recurred at intervals in the theta range in episodes called frames.

Figure 2.5 shows examples of AM pattern in the experiments with New Zealand rabbits, which had been chronically implanted with a square array of 8 $\times$ 8 electrodes (5.6 mm x 5.6 mm) over the visual cortex. All procedures were conducted according to protocol approved by the University of California at Berkeley Animal Use and Care Committee. The rabbits were trained in a classical aversive paradigm to discriminate a reinforced conditioned visual stimulus (CS+) from another not reinforced one (CS-) [1]. The time step to determine each frame in Fig. 2.5 was 31 ms. The diagram

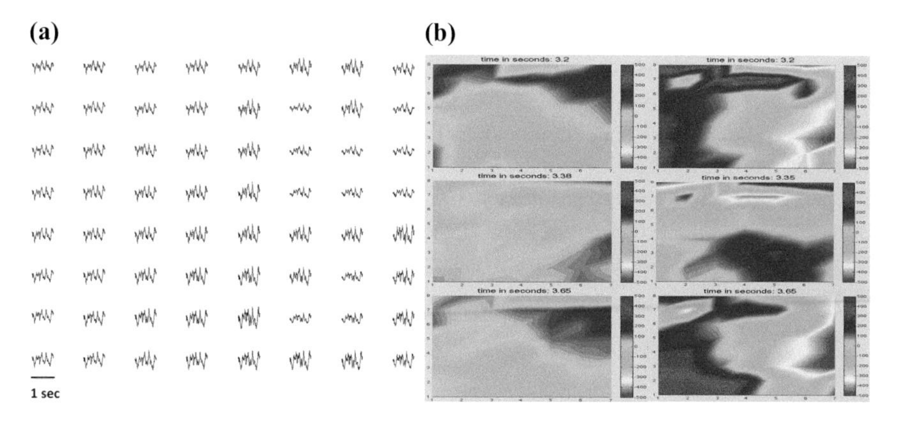

Fig. 2.5 Spatial AM patterns of rabbit experiments with implanted electrodes; **a** an example of the 64 unfiltered ECoG traces in a 1 s window; **b** Temporal evolution of the AM patterns for stimulus A and B in left and right columns, respectively. Stimuli A and B were light flashes, the color of which was reinforced and unreinforced using classical conditioning paradigm

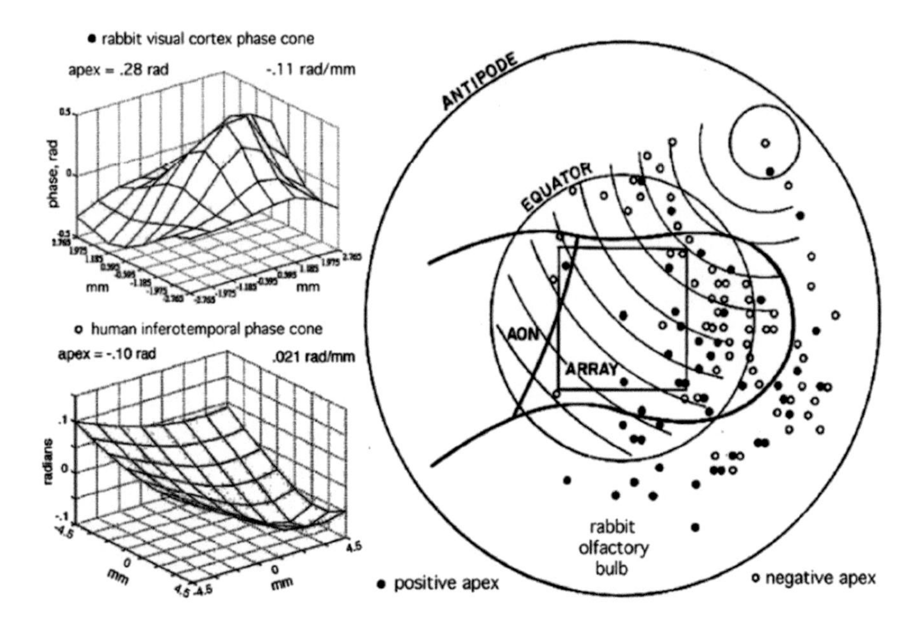

Fig. 2.6 Phase cones in ECoG data. *Left panels* Examples of spatial patterns of the phase of the carrier waveform from rabbit and human, one with an *upward* apex (phase lead, explosion), the other with a *downward* apex (phase lag, implosion). *Right panels* The surface of the rabbit olfactory bulb is *flattened* into a *circle*, with the *square* array outlined in the center. This is an example of iso-phase contours at 0.1 rad intervals; from [35]

illustrates an underlying cycle in AM patterns with a repetition rate of 300–450 ms, corresponding to 2–4Hz in the theta band.

Each AM pattern was accompanied by a spatial pattern of phase modulation (PM) having the form of a cone [35]. The phase is evaluated using Fourier or Hilbert transformation methods and it is independent from the notion of phase in the *phase transition* concept introduced previously in the context of criticality. PM patterns are illustrated in Fig. 2.6, showing that the location and sign of the apices varied randomly. The pair-wise correlations fell with increasing distance between recording sites. We inferred that the interaction strength must also fall with distance [36]. The estimates of diameters showed that cones varied in size and some of them were far larger than the recording array.

2.4 Classification of ECoG and EEG AM Patterns

For quantitative comparison, sequential ECoG patterns were expressed as feature vectors. The vector at each time step specified a state variable that was represented by a point in the brain state space. The Euclidean distance between successive points gave the spatial rate of change. An AM pattern is defined as a set of consecutive frames, which were separated by a distance less than a specific threshold. The center of gravity of the cluster in a frame defined a brain state. Different AM patterns had clusters with centers of gravity that were separated by statistically significant difference in the Euclidean distance between the clusters. A step from one cluster to another defined a state transition, one of Von Neumann's *few short steps* in the action-perception cycle. When two or more clusters of points had been identified, the classification of individual frames was by finding the closest center of gravity by the minimal Euclidean distance. The goodness of classification depended on the number of channels, not on their location. No channel was more or less important than any other; in contrast to the microscopic localization of action potentials, the density of the macroscopic categorizing information was uniformly distributed (Fig. 2.7).

The multidimensional scaling technique of nonlinear mapping [37] projected clusters of points from 64-space into 2-space, optimizing their separation while preserving the relative distances between all of the data points. Two clusters were specified in this example: the 1×192 feature vector from the first three 1×64 feature vectors in the CS+ trials, and the 1×192 feature vector from the first three 1×64 feature vectors in the CS-. The circles representing the spatial standard deviations (SDX) of the clusters were calculated in the display plane; from [38].

In recent years, systematic studies have been conducted to evaluate scalp EEGs in a search for classifiable AM patterns. Feature vectors have been constructed from EEG records of subjects trained in an operant paradigm of reinforcement learning[39]. Pertinent frequency bandwidths have been determined, at which phase locking at steady frequencies emerged. Results by [40] showed the presence of classifiable AM

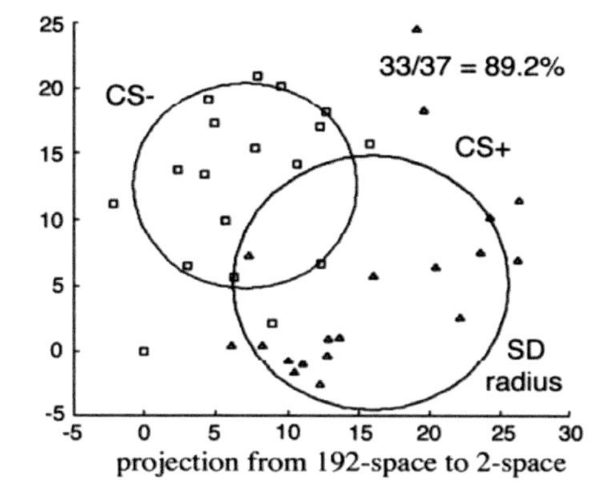

Fig. 2.7 Projection from the 192-dimensional feature space of the 3 main 64-dimensional principal components to the 2-dimensional space. The separation of the CS+ and CS- trials is illustrated by the circles of radii given by the standard deviation drawn around the cluster centers [38]

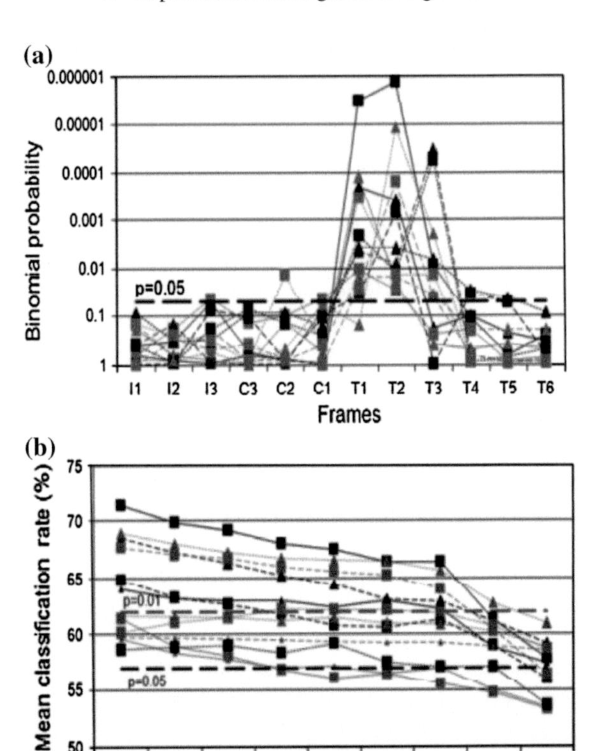

Fig. 2.8 **a** 3 initial frames (I1, I2 and I3), 3 pre-stimulus control frames (C3, C2 and C1), and 6 post-stimulus test frames (T1, T2, T3, T4, T5 and T6) for each of six subjects. **b** The goodness of classification was reduced by removal of channels randomly selected, showing that every electrode contributed equally; from [40], Fig. 3, p. 115

patterns in 64 EEG signals from a standard 10–20 array on the scalp. Classification levels for pre- and post-stimulus periods in the 15–22 Hz pass band were evaluated by the binomial probability of the classification rate of EEG feature vectors extracted from frames determined by high synchrony in the analytic phase: 3 initial frames (I1, I2 and I3), 3 pre-stimulus control frames (C3, C2 and C1), and 6 post-stimulus test frames (T1, T2, T3, T4, T5 and T6) for each of six subjects, see Fig. 2.8.

The goodness of classification depended on the number of channels, not on their location. In contrast to the microscopic localization of action potentials, the density of the macroscopic categorizing information was broadly distributed in the array, as demonstrated by removal of channels. The test of deleting randomly selected channels showed that the correlation distance covered the entire extent of the scalp and underlying neocortex [40].

2.5 Characterization of Synchronization-Desynchronization Transitions in the Cortex

We studied experimental conditions for the transition from a point or nonconvergent attractor to a limit cycle attractor [16]. The search led us to focus on the discontinuities seen in the Analytic Phase (AP), see Fig. 2.9. Differences in the AP were calculated from human EEG in the beta band (12–30 Hz) with 3 mm spacing of 64 electrodes in a linear 189 mm array, digitized at 1 ms intervals [30]. For details of rabbit EEG measurements, see [1]. Globally synchronized beta oscillations were detected in the frames with low variance in phase differences. Dynamic effects with classifiable spatial AM patterns and accompanying PM patterns emerged upon presentation of reinforced stimuli in various sensory systems [41, 42].

We postulated that the attractor landscapes in several sensory systems were simultaneously selected and updated by corollary discharges from the entorhinal cortex in the limbic system during the action-perception cycle [43]. The repetition rate of the patterns was determined by the width of the pass band and not by the center frequency [40, 44, 45]. A potential neural mechanism of the band-pass filter is the negative feedback by inhibitory populations in cortex, combined with Hebbian learning, which amplified the oscillations on reception of input initiated by CS, leading to the formation of Hebbian assemblies [46]. The summation of signals with distributed frequencies gave Rayleigh noise [16, 47] with beat frequencies determined by the width of the distribution at every carrier frequency. Rice predicted the prolongation

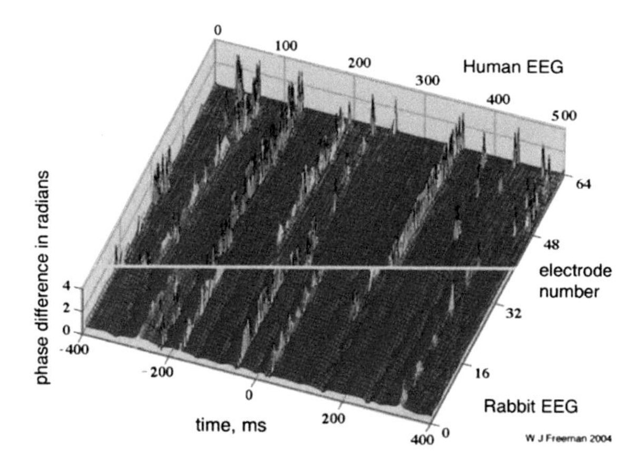

Fig. 2.9 The ECoG and EEG showed that neocortex processed information in frames like a cinema. The search led us to focus on the discontinuities seen in the ECoG between AM patterns. The window of the 8×8 array was 5.6×5.6 mm for the rabbit data (*upper half*). A similar pattern was seen in EEG recordings from a 1×64 curvilinear array 189 mm long, fixed on the scalp of normal human volunteers at rest, eyes open or closed (*lower half*). Globally synchronized beta oscillations were detected in the frames with low variance in phase differences. The similarity in patterns of phase synchrony is extended across a difference in scale of 33:1. The zones of coordinated activity revealed the self-similarity of global dynamics that may form Gestalts (multisensory percepts); [48], cover illustration

of beat intervals by imposing a narrow-band oscillation on the background [44], which was observed in the genesis of wave packets with classifiable AM patterns.

Classifiable AM/PM patterns marked epochs of high order imposed by high interaction intensity and high transmission strength; transitional times between AM/PM patterns marked disorder from weak interactions. We introduce scalar indices by means of which we could calculate and display the degree of order as a function of time. Two scalar indices were developed: the degree of global synchrony of the multichannel ECoG and the goodness of classification of the global AM patterns. In a study designed to search for integration of AM patterns in visual, auditory, somatic, and olfactory areas with the limbic cortex, ECoG potentials were simultaneously recorded from cat or rabbit with electrode arrays that were partitioned among the four major sensory areas plus the entorhinal cortex [49]. Calculation of the analytic phase by means of the Hilbert transform of the signals from the five areas showed 3 to 4 intermittent peaks of macroscopic synchrony in the CS-CR interval. Analysis of the multiple signals with a moving window extracted intermittent AM patterns to which all areas contributed. Removal of the data from any area diminished the goodness of classification of the macroscopic AM patterns.

The succession of bursts of gamma oscillation superimposed on the respiratory wave in the theta range suggested that the olfactory system was bistable [6, 16, 32, 50] as well as quasi-stable. During periods between inhalations the temporal spectrum was $1/f^{\alpha}$, and there were no definable or stable spatial AM/PM patterns. During each inhalation the activity converged to a narrow, steady frequency band carrying fixed AM/PM patterns. This change reflected the transition from a receiving state, which sustained a search trajectory through an unconstrained state space, to a transmitting state, which was tightly constrained by synaptic interactions. The constraints were represented by a low-dimensional attractor governing AM/PM pattern formation. A sharp discontinuity in the analytic frequency demarcated the transition just prior to the appearance of new AM/PM patterns.

2.6 Experimental Observation of Singularity

We found three markers for a singularity in cortical dynamics in ECoG data, which are summarized as follows:

1. The location of the apex of the conic phase modulation pattern.
2. Vortex formation with a center of clockwise or counterclockwise rotation in high-resolution ECoG image sequences.
3. A downward spike (null spike) in the analytic amplitude pattern displays, which was highly localized in time, space, and spectrum.

We predicted that the three points indicating singularity would coincide, i.e., the null spike in the amplitude is accompanied by the temporal discontinuity in the analytic phase and the vortex formation in image sequences [47, 51, 52]. They did so only with exceptionally isolated wave packets. Most windows contained multiple overlapping wave packets with differing carrier frequencies, sizes and center locations. Figure 2.10 shows examples of null spike in the analytic amplitude in rabbit ECoG data, followed by a phase cone emerging from the location of the previous null spike 12 ms later. Figure 2.11 shows the evolution of the amplitude function in rabbit ECoG data [53] between time instances t = 3.56 s and 3.72 s, where dark and light tones indicate low and high amplitudes, respectively. One can observe practically a complete counter-clockwise vortex-like rotation of the analytic amplitude during the given sequence of frames of duration 0.16 s. The presence of vortices in the cortical amplitude patterns has been predicted based on theoretical arguments [54, 55]. Our results show that the synchrony is not completely lost during the transitional period at the null spike; rather it is represented by sequentially appearing semi-localized mesoscopic vortex structures. Thus the described phenomenon is called sequenced synchrony [53]. The experimental observations are often more complex and less clear, as the sequential windows may contain multiple overlapping wave packets with differing carrier frequencies, sizes and center locations.

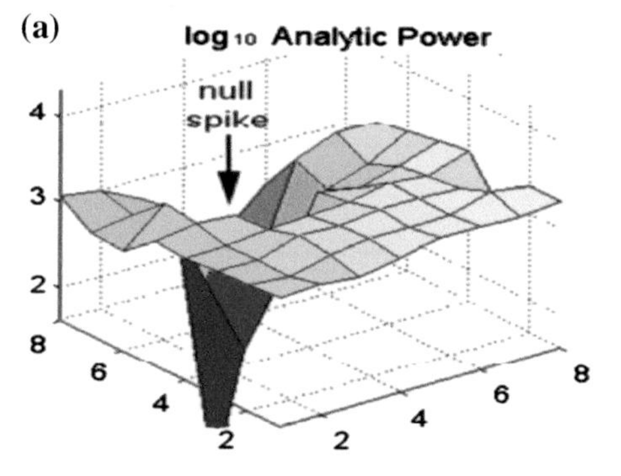

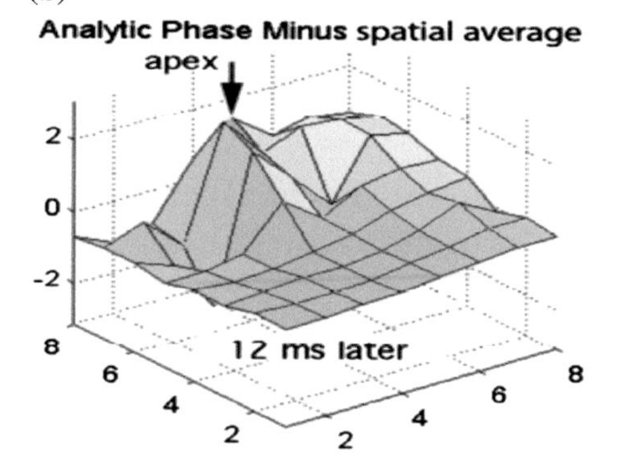

Fig. 2.10 Singular behavior in analytic ECoG signals. **a** Spatial pattern of log10 analytic power in a null spike at the digitizing step of minimum. **b** About 12 ms later (6 digitizing steps) a new phase cone was well established with its maximum at or near the earlier location of the null spike

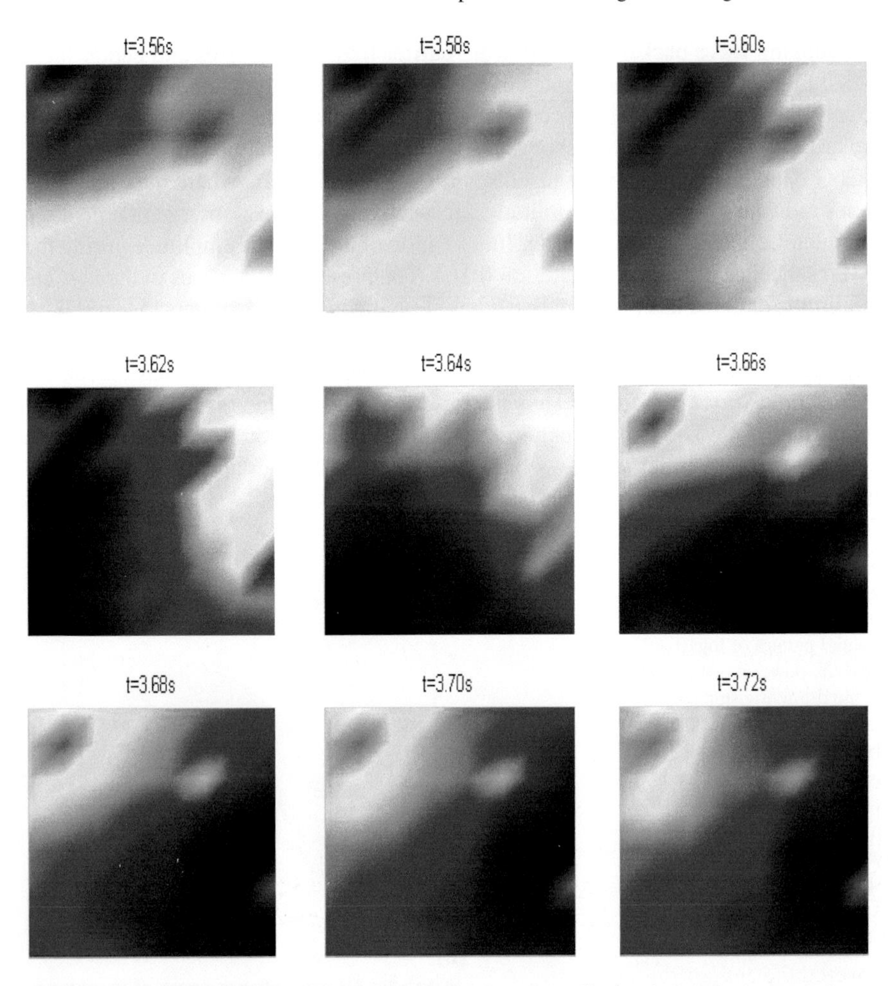

Fig. 2.11 Temporal evolution of the spatial distribution of amplitudes during the sequenced synchrony state. There is an apparent rotation in the vortex counter clockwise from t = 3.56 s–3.72 s

2.7 Transmission of Macroscopic Output by Microscopic Pulses

The phase cones provided no correlates with cognitive content of AM patterns in wave packets; instead they gave us estimates of the size, location, duration, and stationarity of AM patterns. We estimated local correlation distances from the phase of the ECoG from 8×8 square intracranial arrays [36]. The pair-wise correlations fell with increasing distance between recording sites. We inferred that the interaction strength must also fall with distance. We chose the half-power radius of the cone ($cos(\pi/4) \approx 0.7$) as the soft boundary of each neocortical wave packet: $\pi/4$ is divided by the gradient of the cone in rad/mm. The distributions of diameter were skewed and most likely power-law but truncated by spatial filtering from volume conduction. Mean values ranged from 18.6 mm in rabbit to 25.6 mm in human, see Table 2.2 in [48]. The estimates of diameters showed that wave packets were often far larger than the recording arrays. Considering that the density of cortical neurons was $3 \times 105/\text{mm}^2$ areas [56], these diameters would have incorporated roughly 10^7 neurons in a wave packet within a transition time of a quarter cycle (roughly 3–10 ms in the gamma range).

Estimates of the diameters of phase cones at frequencies in the beta range in the CS-CR interval commonly approached the width of the rabbit brain, showing the limitation of the method, which relied on extrapolation of phase gradients measured only within the 8×8 arrays. We sampled over greater distances in search of episodic correlations between individual cortices by using three 4×4 microgrids, one each placed on the visual, auditory, and somatic cortices, a 2×8 array on the entorhinal cortex, plus a pair in the olfactory bulb. What we found was astonishing. Sampling the ECoG from the four 4×4 arrays gave 64×1 feature vectors having at the same frequency components in all cortices carrying classifiable AM patterns with episodic synchrony located over all five areas of recording. Classification was impaired by removing the signals from each of the five participating areas, confirming the uniform density of distributed classificatory information previously shown for each cortex. The recordings from a linear array showed that correlation distances exceeded the length of the array (189 mm).

On the basis of these results we predicted that classifiable AM patterns would also be extracted from scalp EEG records by finding the pertinent frequency and bandwidth, at which time windows of phase locking at constant frequencies emerged. Using those epochs as guidelines, we constructed feature vectors of power [57] in the CS-CR intervals from EEG records of subjects trained in an operant paradigm of reinforcement learning [39]. The prediction was confirmed [40]. by finding classifiable AM pat- terns in 64 EEG signals from a standard 10–20 array on the scalp (Fig. 2.8). The test of deleting randomly selected channels showed that the correlation covered the entire extent of the scalp and underlying neocortex.

We concluded from the multiple power law distributions we encountered in the spatial, temporal and spectral domains that cortical dynamics is scale-free, so that the same neural mechanisms hold for all sizes of population and animal. We focused on the problem of transmission of AM patterns from the olfactory bulb to the prepyriform cortex [46]. Whereas the microscopic sensory information from the receptors to the bulb was by parallel axons in topographic mapping, the bulbar output was carried by a many-to-many pathway, the lateral olfactory tract, which enacted a spatial integral transform of the macroscopic pulse densities carrying the AM pattern information. The result is the Gabor transform that delocalizes the information, so that every column in the cortex has the entire message of the information but at reduced resolution.

Several principles follow. The delocalized AM pattern in the prepyriform cortex is not holographic, or only briefly so for 3–5 cycles, because the carrier oscillation is not monochromatic (a single frequency); it is narrowly distributed about a constant mean value. The width determines the duration of the wave packet at which the frequency distribution determines the phase dispersion that closes the burst [47]. Yet the distribution of the carrier frequencies is so narrow that the coherence lasts 3–5 cycles for the wave packet. The ubiquitous chaotic background is eliminated by the 1/f frequency dispersion within 1 cycle, which cleans the AM pattern. Since the AM pattern is the memory and the rest is clutter, the Gabor operation solves for cortex the figure-ground problem.

The prepyriform cortex sustains the cleaned, delocalized oscillatory AM pattern by the negative feedback interaction between the small superficial pyramidal excitatory cells and the stellate inhibitory cells in Layer II. However, it is not self-organized as it is in the bulb. The evidence is that the wave packets of the prepyriform cortex do not have phase cones, in which the location of the apex is a random variable, and the slope is a function of the conduction velocity of axons in the cortex. The wave is a linear front always anterior to posterior at a gradient determined by the axons in the lateral olfactory tract. The bulb is the driver; the cortex is the driven.

The larger deep pyramidal cells in Layer III sum the local pulse activity in the cortical neuropil and fire in proportion to the local field potential, as shown by calculation of the probability of pulse firing conditional on LFP amplitude. The collective ensemble of pulses on axons of the deep pyramidal cells carries the same AM pattern amplitude and memory to all targets of prepyriform transmission at a resolution proportional to the from of the whole for each target.

In conclusion, we point out the salient points that while the bulb transforms the microscopic sensory information into a macroscopic field of wave density, the prepyriform cortex reads the AM pattern and sends to other cortices in the same microscopic firing code as used by the primary sensory cortices [58]. The implication from the olfactory system is that the phase transition requires that neural masses work in pairs, one to construct an macroscopic memory and the second to read it and concert it into a microscopic form that sensory and motor systems can use in the same code as sensory input. The bulb and prepyriform are a prototype; comparable pairs may be sought in entorhinal cortex and hippocampus,V1 and V2, V1 and the pulvinar, and similar pairs in other sensory systems.

References

1. Barrie JM, Freeman WJ, Lenhart M (1996) Modulation by discriminative training of spatial patterns of gamma EEG amplitude and phase in neocortex of rabbits. J Neurophysiol 76:520–539
2. Freeman WJ, Holmes MD, West GA, Vanhatalo S (2006) Fine spatiotemporal structure of phase in human intracranial EEG. Clin Neurophysiol 117:1228–1243
3. Menon V, Freeman WJ, Cutillo BA, Desmond JE, Ward MF, Bressler SL, Laxer KD, Barbaro N, Gevins AS (1996) Spatio-temporal correlations in human gamma band electrocorticograms. Electroencephalogr Clin Neurophysiol 98(2):89–102
4. Freeman WJ (1995) The Hebbian paradigm reintegrated: local reverberations as internal representations. Behav Brain Sci 18(4):631–631
5. Freeman WJ, (2000/2006) Neurodynamics. An Exploration of Mesoscopic Brain Dynamics. Springer, London. Electronic version, http://sulcus.berkeley.edu/

6. Freeman WJ, Burke BC, Holmes MD (2003) Aperiodic phase re-setting in scalp EEG of beta-gamma oscillations by state transitions at alpha-theta rates. Hum Brain Mapp 19(4):248–272
7. Liley DT, Alexander DM, Wright JJ, Aldous MD (1999) Alpha rhythm emerges from large-scale networks of realistically coupled multicompartmental model cortical neurons. Netw: Comput Neural Syst 10(1):79–92
8. Nunez PL, Cutillo BA (eds) (1995) Neocortical dynamics and human EEG rhythms. Oxford University Press, New York
9. Wright JJ, Liley DTJ (1996) Dynamics of the brain at global and microscopic scales. Neural Netw EEG, Behav Brain Sci 19(1996):285–320
10. Tallon-Baudry C, Bertrand O, Delpuech C, Pernier J (1996) Stimulus-specificity of phase-locked and non phase-locked 40-Hz visual responses in human. J Neurosci 16:4240–4249
11. Tallon-Baudry C, Bertrand O, Peronnet F, Pernier J (1998) Induced gamma-band activity during the delay of a visual short-term memory task in humans. J Neurosci 18:4244–4254
12. Muller MM, Bosch J, Elbert T, Kreiter A, Valdes Sosa M, Valdes Sosa P, Rockstroh B (1996) Visually induced gamma band responses in human EEG—a link to animal studies. Exp Brain Res 112:96–112
13. Rodriguez E, George N, Lachaux J-P, Martinerie J, Renault B, Varela F (1999) Perception's shadow: long-distance synchronization of human brain activity. Nature 397:430–433
14. Miltner WHR, Braun C, Matthias A, Witte H, Taub E (1999) Coherence of gamma-band EEG activity as a basis for associative learning. Nature 397(6718):434–436
15. Milner B (1966) Amnesia following operation on the temporal lobes. In: Whitty CWM, Zangwill OM (eds) Amnesia. Butterworths, London, pp 109–133
16. Freeman WJ (1975/2004) Mass action in the nervous system. Academic Press, New York. Electronic version 2004—http://sulcus.berkeley.edu/MANSWWW/MANSWWW.html
17. von der Malsburg C (1983) How are nervous structures organized? In: Basar E, Flohr H, Haken H, Mandell AJ (eds) Synergetics of the Brain. Springer, Berlin, pp 238–249
18. Singer W, Gray CM (1995) Visual feature integration and the temporal correlation hypothesis. Annu Rev Neurosci 18:555–586
19. Ramon C, Freeman WJ, Holmes M, Ishimaru A, Haueisen J, Schimpf PH, Rezvanian E (2009) Similarities between simulated spatial spectra of scalp EEG MEG Structural MRI. Brain topogr 22(3):191–196
20. Ramon C, Schimpf PH, Haueisen J (2006) Influence of head models on EEG simulations and inverse source localizations. Biomed Eng Online 5(1):10
21. Haueisen J, Tuch DS, Ramon C, Schimpf PH, Wedeen VJ, George JS, Belliveau JW (2002) The influence of brain tissue anisotropy on human EEG and MEG. Neuroimage 15(1):159–166
22. Ramon C, Gehin C, Schmitt PM (2006) Interface pressure monitoring for a secured instrumented childbirth. In: 28th Annual International Conference of the IEEE Engineering in Medicine and Biology Society EMBS'06, p 3186–3189
23. Geddes LA, Baker LE (1967) The specific resistance of biological materials. A compendium of data for the biomedical engineer and physiologist. Med Biol Eng 5(3):271–293
24. Foster KR, Schwan HP (1989) Dielectric properties of tissues and biological materials: a critical review. CRC Crit Rev Biomed Eng 17:25–104
25. Gabriel S, Lau RW, Gabriel C (1996) The dielectric properties of biological tissues: II measurements in the frequency range 10 Hz to 20 GHz. Phys Med Biol 41(11):2251
26. Schimpf PH (2007) Application of quasi-static magnetic reciprocity to finite element models of the meg lead-field. IEEE Trans Biomed Eng 54(11):2082–2088
27. Freeman WJ (2001) How brains make up their minds. Columbia University Press, New York
28. Schroeder M (1991) Fractals, chaos power laws. WH Freeman Publishers, New York
29. Freeman WJ, Zhai J (2009) Simulated power spectral density (PSD) of background electrocorticogram (ECoG). Cogn Neurodyn 3(1):97–103
30. Freeman WJ (2004) Origin, structure, and role of background EEG activity. Part 1 Analytic amplitude. Clin Neurophysiol 115:2077–2088
31. Hartline HK, Ratliff F (1958) Spatial summation of inhibitory influences in the eye of Limulus and the mutual interaction of receptor units. J Gen Physiol 41:1049–1066

32. Freyer F, Aquino K, Robinson PA, Ritter P, Breakspear M (2009) Bistability and non-Gaussian fluctuations in spontaneous cortical activity. J Neurosci 29(26):8512–8524
33. Pikovsky A, Rosenblum M, Kurths J (2001) Synchronization a universal concept non-linear science. Cambridge University Press, Cambridge
34. Freeman WJ, Burke BC (2003) A neurobiological theory of meaning in perception. Part 4. Multicortical patterns of amplitude modulation in gamma EEG. Int J Bifurc Chaos 13:2857–2866
35. Freeman WJ, Baird B (1987) Relation of olfactory EEG to behavior: spatial analysis. Behav Neurosci 101(3):393
36. Freeman WJ, Barrie JM (2000) Analysis of spatial patterns of phase in neocortical gamma EEGs in rabbit. J Neurophysiol 84(3):1266–1278
37. Sammon JW (1969) A nonlinear mapping for data structure analysis. IEEE Trans Comput, C-18:401–409
38. Freeman WJ (2005) Origin, structure, and role of background EEG activity. Part 3. Neural frame classification. Clin Neurophysiol 116(5):1118–1129
39. Pockett S, Bold GEJ, Freeman WJ (2009) EEG synchrony during a perceptual-cognitive task: widespread phase synchrony at all frequencies. Clin Neurophysiol 120:695–708
40. Ruiz Y, Pockett S, Freeman WJ, Gonzales E, Guang Li (2010) A method to study global spatial patterns related to sensory perception in scalp EEG. J Neurosci Methods 191:110–118
41. Ohl FW, Scheich H, Freeman WJ (2001) Change in pattern of ongoing cortical activity with auditory category learning. Nature 412:733–736
42. Kozma R, Freeman WJ (2002) Classification of EEG patterns using nonlinear neurodynamics and chaos. Neurocomputing 44–46:1107–1112
43. Kay LM, Freeman WJ (1998) Bidirectional processing in the olfactory-limbic axis during olfactory behavior. Behav Neurosci 112:541–553
44. Rice SO (1950) Mathematical analysis of random noise and appendixes., Technical Publications Monograph B-1589Bell Telephone Labs Inc, New York
45. Freeman WJ, Ahlfors SM, Menon V (2009) Combining EEG, MEG and fMRI signals to characterize mesoscopic patterns of brain activity related to cognition. Special issue (Lorig TS (ed)). Int J Psychophysiol 73(1):43–52
46. Freeman WJ, Quian Quiroga R (2013) Imaging brain function with EEG: advanced temporal and spatial analysis of electroencephalographic and electrocorticographic signals. Springer, New York
47. Freeman WJ (2009) Deep analysis of perception through dynamic structures that emerge in cortical activity from self-regulated noise. Cogn Neurodyn 3(1):105–116
48. Freeman WJ (2004) Origin, structure, and role of background EEG activity. Part 2 Analytic phase. Clin Neurophysiol 115:2089–2107
49. Freeman WJ, Gaál G, Jorsten R (2003) A neurobiological theory of meaning in perception part III: multiple cortical areas synchronize without loss of local autonomy. Int J Bifurc Chaos 13(10):2845–2856
50. Skarda CA, Freeman WJ (1987) How brains make chaos in order to make sense of the world. Behav Brain Sci 10:161–195
51. Freeman WJ, Kozma R (2010) Freeman's mass action. Scholarpedia 5(1):8040
52. Freeman WJ, Vitiello G (2010) Vortices in brain waves. Int J Mod Phy B 24(17):3269–3295
53. Kozma R, Freeman WJ (2008) Intermittent spatio-temporal de-synchronization and sequenced synchrony in ECoG signals. Chaos 18:037131
54. Freeman WJ, Rogers LJ (2002) Fine temporal resolution of analytic phase reveals episodic synchronization by state transitions in gamma EEG. J Neurophysiol 87:937–945
55. Freeman WJ (2007) Proposed cortical "shutter" mechanism in cinematographic perception. In: Perlovsky L, Kozma R (eds) Neurodynamics of cognition and consciousness. Springer, Heidelberg, pp 11–38
56. Braitenberg V, Schuz A (1998) Cortex: statistics and geometry of neuronal connectivity, 2nd edn. Springer, Berlin

57. Capolupo A, Freeman WJ, Vitiello G (2013) Dissipation of 'dark energy' by cortex in knowledge retrieval. Phys Life Rev, Online. doi:10.1016/j.plrev.2013.01.001
58. Freeman WJ (2015) Perspectives: mechanism and significance of global coherence in scalp EEG. In: Buzsaki G, Freeman WJ, (eds)) Current opinion in neurobiology 31: brain rhythms and coordination, pp 199–207

Chapter 3
Interpretation of Experimental Results As Cortical Phase Transitions

3.1 Theoretical Approaches to Nonlinear Cortical Dynamics

Brain activity can be interpreted in terms of dynamic system theory [1–3], in particular using models of transient dynamics [4–6]. Some models utilize encoding in complex cycles and chaotic attractors [7–9]. A hierarchical approach to neural dynamics is formulated as the Freeman K model [10, 11], which is based on studying olfaction, and later generalized to other sensory systems. In the intact olfactory system, the oscillations in bulbar and cortical ECoG are chaotic due to the interaction of oscillatory brain regions with incommensurate frequencies [12]. Complex partial seizures may arise if the balance between the oscillatory components is broken. The long feedback connections have propagation delays that approach the half-cycle duration of the frequencies of the oscillatory components. These conditions yield sustained aperiodic oscillations that imply the existence of a nonconvergent attractor in the rest state and in the absence of sustained input from the olfactory receptors.

Multiple spatial AM patterns imply the existence of intermittent attractor landscape. The examples of the ECoG shown in Fig. 3.1 (right) were from four states: coma induced by deep anesthesia; rest; normal intentional behavior; and complex partial seizure. From modeling with ODE we inferred that the lower pair shows the transition from a zero point attractor in deep anesthesia to a chaotic attractor on recovery to rest. With sustained arousal and with each inhalation, we inferred that a landscape emerged of limit cycle attractors [12].

Classifiable AM patterns emerge only during inhalations. From the dependence of the AM patterns on the presence of one of the reinforced conditioned stimuli (CS+), we inferred that each basin was established over successive sniffs of the same reinforced conditional stimulus (CS+) given to randomly varying receptors across sniffs. Cumulatively, over multiple sniffs, a Hebbian nerve cell assembly has been formed by strengthening the connections between pairs of excitatory neurons co-excited by CS+ [13], and by weakening the connections among co-excited neurons by CS−. Thereafter excitation of any subset in the assembly by the CS+ acti-

R. Kozma and W.J. Freeman, *Cognitive Phase Transitions in the Cerebral Cortex – Enhancing the Neuron Doctrine by Modeling Neural Fields*,
Studies in Systems, Decision and Control 39, DOI 10.1007/978-3-319-24406-8_3

vated the entire mesoscopic assembly. The Hebbian assembly generalized the bulbar response to the class of receptors that had been activated under reinforcement during training [14].

We implemented this interpretation by approximating the infinite-dimensional cortical state by the experimental 64×1 column vector. The moving tip of the vector inscribed a trajectory in 64-space. Transition from a more random state to a more ordered state occurred when the trajectory was confined in proximity to a limit cycle attractor. Emergence of the landscape required two conditions: arousal, and an act of observation (sniff) bringing a surge of sensory input. Selection of a basin of attraction required presentation of a stimulus that a subject had been trained to discriminate. The emergence of stable patterns in the phase and amplitude of the multichannel ECoG demonstrated convergence to the attractor. We further postulated that an assembly with its basin of attraction existed in the olfactory bulb for each class of odorant that a subject had learned to categorize, and that a conditioned stimulus during a sniff selected the corresponding attractor by directing the high-dimensional search trajectory between frames into the basin of attraction. The collapse of the entire landscape during exhalation frees the system for the next sniff.

The examples of the ECoG shown in Fig. 3.1 (right) were from four states: coma induced by deep anesthesia; rest; normal intentional behavior; and complex partial

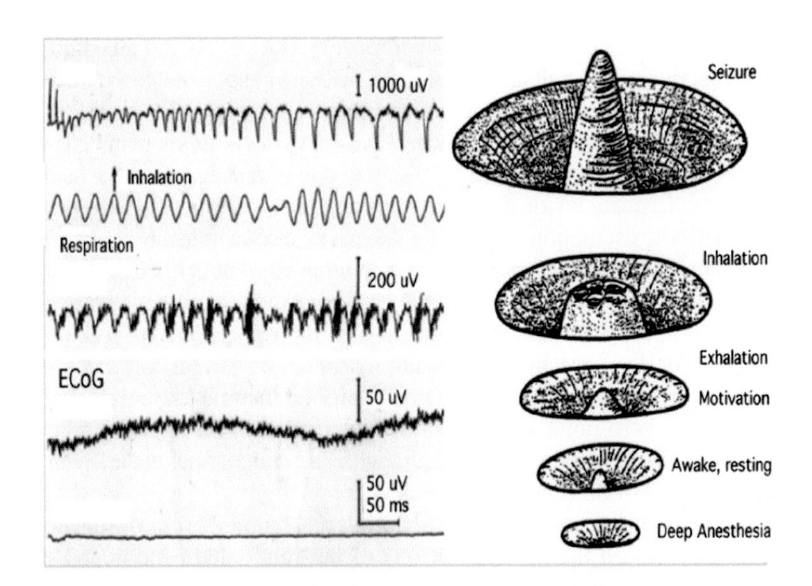

Fig. 3.1 Illustration of principles of chaotic brain dynamics; *Left* examples of olfactory ECoG in states of deep anesthesia, normal operation, and seizure. *Right* hypothetical phase portraits representing states in the olfactory system, ranging from an open loop, zero point attractor under deep anesthesia upward into noise at rest, metastability in arousal, up to intense activity in complex partial seizure. In the normal state of intentional behavior inhalation results in the emergence of a landscape of learned limit cycle attractors, one of which may be selected by an odorant input that places the system in its basin. Alternatively, the trajectory may fall into the chaotic well. On exhalation, the learned attractor landscape vanishes; from [12]

seizure. With sustained arousal and with each inhalation a landscape of limit cycle attractors emerged [12]. In the normal state of intentional behavior inhalation results in the emergence of a landscape of learned limit cycle attractors, one of which may be selected by an odorant input that places the system in its basin. Alternatively, the trajectory may fall into the chaotic well. This appears to occur on about 10 % of control inhalations and about 40 % of the test odor inhalations after completion of training, as well as reliably with novel odorants [15].

3.2 Scales of Representation: Micro-, Meso-, and Macroscopic Levels

We made our observations of brain activity and represented them in models as state variables at three differing scales in time and space: micro-, meso-, macro-. In prior reports we defined the three scales in terms of the method of observation. We observed microscopic events with microelectrodes as action potentials (spikes, units, pulses) and represented the spike trains as point processes. We observed mesoscopic processes with intracranial macroelectrodes: ECoG from cortical surfaces and LFP (local field potentials) from depth electrodes, integrating signals for about 10^4 neural units. We observed macroscopic behaviors noninvasively using large-scale EEG, MEG and fMRI and represented them as images [16].

Here we redefine the scales by including the behavioral correlates of the neural activity. Microscopic activity consists of axonal action potentials that are directly correlated with sensory stimuli and motor actions [17]. Macroscopic data are derived from measurements of extracellularly recorded dendritic potentials in all forms, which are related to the perceptions of stimuli and the intentions to act, not to the stimuli or acts. The data lack invariance with fixed stimuli and actions but change with learning. Furthermore, each measurement is determined by summation of simultaneous contributions from many neurons in each local neighborhood, that is, a population. Mesoscopic activity consists of cell spikes that are related to abstract concepts, which are typically unique to the individual subjects. Examples are feature detector neurons [18], mirror neurons [19], and face neurons [20]. The fact that such neurons can be found at all among the 10^5 neurons in each cubic mm implies that they must be numerous and firing in parallel. They are mesoscopic, because they can only be observed with microelectrodes, yet they have behavioral correlates that imply some degrees of abstraction and generalization from the specificities of sensation and reflex action.

We interpret experimental data obtained by ECoG and EEG arrays as the manifestation of transient spatio-temporal brain processes, whereas the brain dynamics is intermittently captured in a localized attractor basin, as evidenced by the convergence to metastable AM/PM patterns with steady carrier frequencies. The convergence was marked by discontinuities in the phase and amplitude by high ECoG resolution in both temporal and spatial dimensions at pattern onsets. Termination of the patterns was by the disappearance of the landscape. We observe an extended dwell time in the basin, which is much longer than the rapid transition time from one attractor basin to the other. The discontinuities are critically important for the successful transmission of the contents of the wave packets by way of synchronization across large cortical areas.

3.3 Cinematic Theory of Cortical Phase Transitions

The ECoG and EEG showed that neocortex processed information in frames like a cinema. The metastable AM patterns manifest the frames, and the phase transitions provide the shutter from each frame to the next. Moving from one metastable pattern to the other corresponds to successive images in a movie. The cinematic theory of cognition [21–24] seems to be compatible with the mathematical concept of heteroclinic cycles [25, 26]. In networks with competing components, heteroclinic channels connect metastable states and produce transient brain dynamics. In an alternative model, Tsuda described the dynamic brain trajectory as cruising through high-dimensional state space in chaotic itinerancy [27]. He conceived that Milnor attractors dissolved into attractor ruins, which forestalled capture into rigid stability.

In this work we follow Haken's synergetic approach to information processing [28] to establish a modeling framework for the cinematic concept. Haken proposed that state transitions are essential for information transfer between hierarchical levels, by which a collection of particles create an order parameter and in circular causality enslaves the activity of the particles. Metastability of large-scale global cortical dynamics has been described in [29]. Kelso and Engstrom [30] studied coordination dynamics in the cerebrum during rhythmic finger movements and accompanying EEG waves, leading to the theory of the metastable brain [31].

We adapted the thermodynamic concept of phase [6, 32–34]. Note that this concept of phase is related to the phase transition, and it is completely different from the phase obtained by Hilbert or Fourier transforms and incorporated in PM patterns. We define two states of the cortex: a disordered receiving phase carrying microscopic information, and an ordered transmitting phase carrying macroscopic information. In this model we observe two phase transitions, one abrupt on the initiation of an AM pattern by condensation from disorder to an ordered state, like the condensation of vapor to liquid [35]. The other transition is more gradual by progressive dissolution of order in returning to the high-dimensional, disordered phase (evaporation). The alternation was revealed most clearly by plotting the successive differences in analytic phase by the Hilbert transform of the 64 ECoG signals after band-pass filtering, see Fig. 3.2.

We observed that the initial site where non-homogeneous condensation starts (the phase cone apex) is not conditioned by the incoming stimulus, but is randomly determined by the concurrence of a number of local conditions, such as where the null spike is lowest and the background input is highest, in which the cortex finds itself at the time of the transition. The null spike appears in the band-pass filtered noise

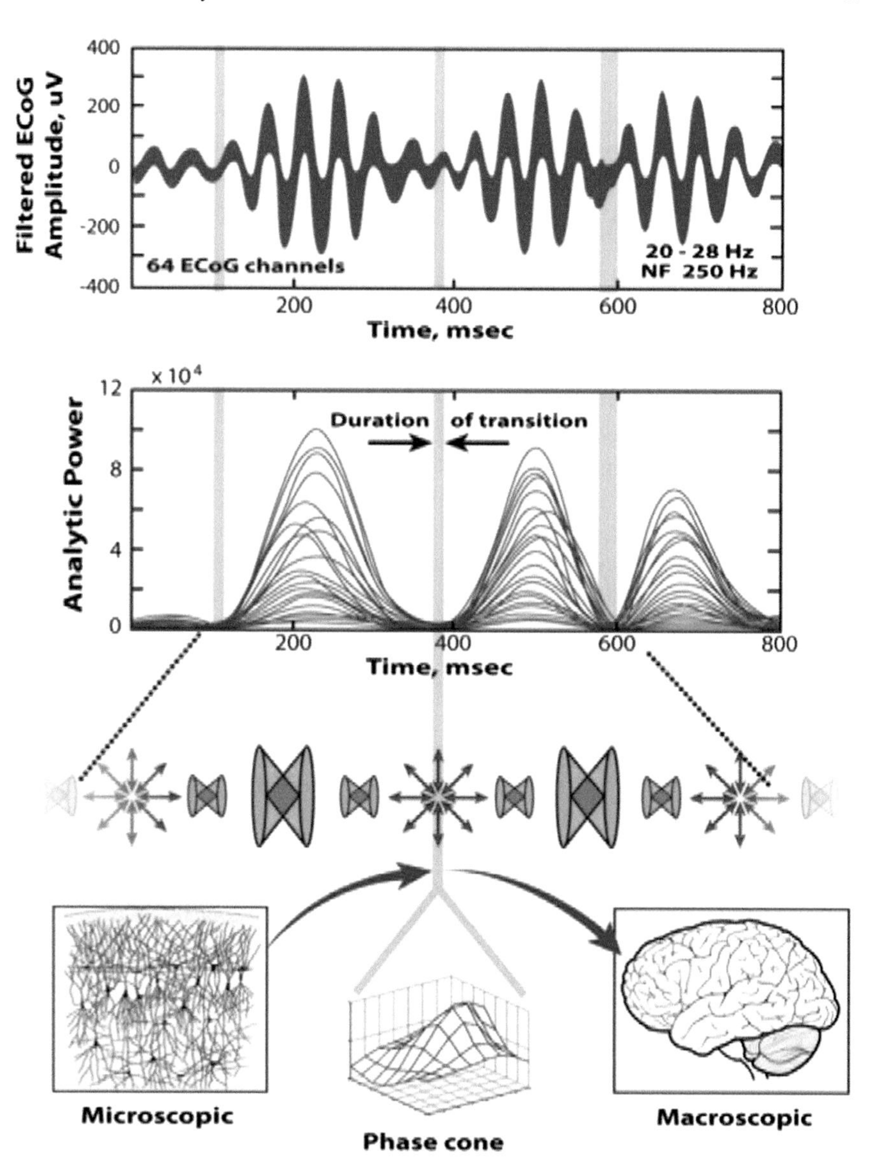

Fig. 3.2 Illustration of the cognitive cycle; *top* 64 superimposed band-pass filtered ECoG signals. The 64 analytic amplitudes show reduction during the intermittent singularity (*blue band*). These are the null spikes (within vertical *blue bars*), which are spatially and temporally localized. *Bottom* During the singularity, phase cones convey the transition from microscopic disorder to macroscopic order (illustration by Chris Gralapp)

activity and can be conceived as a shutter that blanks the intrinsic background ECoG. When the order parameter goes to zero, the microscopic activity of the background state does not decrease but it becomes disordered. In such a state of very low analytic amplitude, the analytic phase is undefined, and the system, under the incoming weak sensory input, may re-set the background activity in a new AM frame, thus completing a phase transition. The aperiodic shutter provides opportunity for the emergence of phase transitions. The analytic amplitude decrease repeats in the theta or alpha range, independently of the repetitive sampling of the environment by limbic input.

The presence of incipient, smaller phase cones during the existence of metastable AM patterns [36] suggests that the ECoG intermittently resembled avalanches of various sizes that maintain the state of self-organized criticality (SOC) [37, 38]. Avalanche dynamics has been identified in neural activity [39, 40]. Avalanches of neuronal activity occur spontaneously in the superficial layers of the cerebral cortex under a variety of experimental conditions. The power law distribution of avalanche sizes suggests that the tissue samples are in a dynamic state of criticality, possibly attributable to branching processes [41]. Incipient phase cones manifest the release of energy in weak bursts that dissipate the energy of the cortex released by bombardment with irrelevant background input [35]. Only the relevant CSs can cross the threshold leading to condensation.

From the viewpoint of cinematic theory of cognition, the onset of null spikes and singular space-time dynamics are of critical importance. Occasionally, under the impact of a CS that activates a Hebbian assembly, a phase cone grows large enough and persists long enough to induce a phase transition through a spatio-temporal singularity. Such effect is beyond the scale-free dynamics predicted by SOC, as the phase transition produces significant outliers from SOC statistics. Deviations from SOC have been recently identified as *Dragon Kings* [42], and ECoG dynamics may provide additional examples of such crisis events [23, 43–45]. Exceptional, singular events in the ECoG are manifested by apices of large-scale phase cones, which accompany AM patterns carrying cognitive contents. We postulated that creation of a classifiable AM pattern required transition energy, which came from the amplification of gamma oscillation by a conditioned stimulus that enabled the onset of the phase transition. Most importantly we infer that only CSs can elicit phase transitions. The phase transition commits the neuropil to a trajectory for the duration of the burst at a high cost of energy and is not to be undertaken lightly. In Chap. 10 we elaborate on structural properties of the neuropil, which are especially favorable for rapid phase transitions. Such properties include the presence of *Rich Club* connectivity [46], hierarchical cortical structures evolving from pioneer neurons [47], and the possible presence of scale-free connection length distributions, which has been a highly debated issue for more than a decade.

3.4 Characterization of Phase Transitions

We postulate that phase transition would require us to demonstrate six features [48]:

1. Singularity in cortical dynamics;
2. Stable, self-regulated state of criticality;
3. Symmetry breaking;
4. Transition energy;
5. Change in the order parameter from zero to a finite value; and
6. Major increase in correlation distance in the ordered state.

3.4.1 Critical State

We observe that the initial site where non-homogeneous condensation starts (the phase cone apex) is not conditioned by the incoming stimulus, but is randomly determined by the concurrence of a number of local conditions, such as where the null spike is lowest and the background input is highest, in which the cortex finds itself at the transition process time. The apex is never initiated within frames (in the broken symmetry phase or ordered region), but between frames (during phase transitions), as it is indeed predicted by the dissipative model (vortices occur during the critical regime of phase transitions).

3.4.2 Singular Dynamics

The null spike appears in the band pass filtered brown noise activity and can be conceived as a shutter that blanks the intrinsic background ECoG. When the order parameter goes to zero the microscopic activity (of the background state) does not decrease but it becomes disordered (fully symmetric). In such a state of very low analytic amplitude, the analytic phase is undefined, as it is indeed at the center line of the vortex core, and the system, under the incoming weak sensory input, may re-set the background activity in a new AM frame, if any, formed by reorganizing the existing activity. The analytic amplitude decrease repeats in the theta or alpha range, independently of the repetitive sampling of the environment by limbic input. Consistently with observations, in the dissipative model the reduction in activity constitutes a singularity in the dynamics at which the phase is undefined. The aperiodic shutter allows opportunities for phase transitions.

The receiving phase is characterized by oscillations at all frequencies with self-similarity of the waveforms of the ECoG in all pass bands in the beta-gamma range [35], as predicted from the $1/f$ power-law distributions of the PSDT and PSDX [36, 49], Part 4. Each frequency in the PSDT from the FFT had a phase surface from the 64 ECoG signals. We fitted a phase cone to the surface at each time step to get the location and sign of the apex and estimated the goodness of fit by the size of the residuals. The durations were estimated by counting the number of time steps across which the same apex persisted. The distributions were power-law, and they varied with the size of the window used to calculate the FFT, implying that they were fractal. The appearance suggested that the ECoG resembled a mist, in which droplets of water continually condensed and evaporated in criticality [35, 50], only occasionally growing large enough and persisting long enough to transit unequivocally to the condensed phase and drop as rain. The mini-phase cones also resemble the avalanches that maintain the critical angle of a sand pile [38] in self-organized criticality. The exceptional events in the ECoG are the wave packets carrying cognitive contents; the aim of this research is to explain the neural mechanisms that construct them [51].

3.4.3 Symmetry Breaking

The receiving phase lacked AM patterns and long-lasting phase cones, and the spectra were power-law. The featureless background ECoG was similar under shifts in time, location and orientation; by that criterion it was symmetric. Emergence of structure following a null spike broke the symmetry.

3.4.4 Transition Energy

We postulated that creation of a classifiable AM pattern required transition energy, which came from the amplification of gamma oscillation by a Hebbian assembly.

3.4.5 Zero Order Parameter

The degree of order continually fluctuated in every pass band of the ECoG, as reflected in the envelopes of amplitudes, the intermittent stationarity of the carrier frequency, and the degree of synchrony. The repetition rate of the down spikes was solely determined by the width of the pass band and not by the center frequency [52–54]. We postulated that the neural mechanism of the band pass filter was the negative feedback among $K I_e$ and $K I_i$ populations in cortex, combined with Hebbian learning, which amplified the oscillations on reception of input initiated by CS. We measured the distribution of frequencies in the carrier from the SDX(t) of the 64 estimates of the carrier frequencies of wave packets. The summation of signals with distributed frequencies gave Rayleigh noise [10, p. 148] with beat frequencies determined by the width of the distribution at every carrier frequency. Rice predicted the prolongation of beat intervals by imposing a narrow band oscillation on the background, which was observed in the genesis of wave packets with classifiable AM patterns. Of particular interest was the null spike during a downbeat, in which the ECoG approached zero for a very brief moment in a point on the cortical surface, when the order parameter virtually vanished, demarcating a singularity.

3.4.6 Correlation Length Divergence

According to statistical theory of criticality, the correlation length diverges as the system approaches critical states. The experimentally observed increase of correlation length has confirmed these predictions [48, 55].

Next, we develop a mathematical theory of phase transitions based on random graph theory (RGT) to describe the observed intermittent spatio-temporal oscillations in terms of the collective activity of large-scale neural populations.

References

1. Freeman WJ (1991) The physiology of perception. Sci Am 264(2):78–85
2. Hoppensteadt FC, Izhkevich EM (1998) Thalamo-cortical interactions modeled by weakly connected oscillators: could the brain use FM radio principles? BioSystem 48:85–94
3. Korn H, Faure P (2003) Is there chaos in the brain? II. Experimental evidence and related models. C R Biol 326:787–840
4. Rabinovich MI, Abarbanel HDI (1998) The role of chaos in neural systems. Neuroscience 87(1):5–14
5. Kaneko K, Tsuda I (2001) Complex systems: chaos and beyond. A constructive approach with applications in life sciences. Springer, Berlin
6. Steyn-Ross DA, Steyn-Ross ML (eds) (2010) Modeling phase transitions in the brain, Springer series in computational neuroscience, vol 4. Springer, Berlin
7. Aihara K, Takabe T, Toyoda M (1990) Chaotic neural network. Phys Lett A 144:333–340
8. Andreyev YV, Dimitriev AS, Kuminov DA (1996) 1-D maps, chaos and neural networks for information processing. Int J Bifurc Chaos 6(4):627–646
9. Borisyuk RM, Borisyuk GN (1997) Information coding on the basis of synchronization neuronal activity. Biosystems 40(1–2):3–10
10. Freeman WJ (1975/2004) Mass action in the nervous system. Academic, New York. Electronic version 2004. http://sulcus.berkeley.edu/MANSWWW/MANSWWW.html
11. Freeman WJ (2001) How brains make up their minds. Columbia UP, New York
12. Skarda CA, Freeman WJ (1987) How brains make chaos in order to make sense of the world. Behav Brain Sci 10:161–195
13. Emery JD, Freeman WJ (1969) Pattern analysis of cortical evoked potential parameters during attention changes. Physiol Behav 4:67–77
14. Freeman WJ (1979) Nonlinear dynamics of paleocortex manifested in the olfactory EEG. Biol Cybern 35:21–37
15. Freeman WJ, Baird B (1987) Relation of olfactory EEG to behavior: spatial analysis. Behav Neurosci 101(3):393
16. Freeman WJ, Ahlfors SM, Menon V (2009) Combining EEG, MEG and fMRI signals to characterize mesoscopic patterns of brain activity related to cognition. Special Issue (Lorig TS ed) Int J Psychophysiol 73(1): 43–52
17. Hartline HK, Ratliff F (1958) Spatial summation of inhibitory influences in the eye of Limulus and the mutual interaction of receptor units. J Gen Physiol 41:1049–1066

18. Singer W, Gray CM (1995) Visual feature integration and the temporal correlation hypothesis. Annu Rev Neurosci 18:555–586
19. Rizzolatti G, Craighero L (2004) The mirror-neuron system. Annu Rev Neurosci 27:169–192
20. Quian Quiroga R, Panzeri S (2009) Extracting information from neuronal populations: information theory and decoding approaches. Nat Rev Neurosci 10:173–185
21. Freeman WJ (2007) Proposed cortical "shutter" mechanism in cinematographic perception. In: Perlovsky L, Kozma R (eds) Neurodynamics of cognition and consciousness. Springer, Heidelberg, pp 11–38
22. Freeman WJ, Kozma R, Vitiello G (2012) Adaptation of the generalized Carnot cycle to describe thermodynamics of cerebral cortex. In: The 2012 international joint conference on neural networks (IJCNN). IEEE, pp 1–8
23. Kozma R, Puljic M, Freeman WJ (2012) Thermodynamic model of criticality in the cortex based on EEG/ECoG data. arXiv preprint arXiv:1206.1108
24. Freeman WJ, Quian Quiroga R (2013) Imaging brain function with EEG: advanced temporal and spatial analysis of electroencephalographic and electrocorticographic signals. Springer, New York
25. Rabinovich MI, Friston KJ, Varona P (eds) (2012) Principles of brain dynamics. MIT Press, Cambridge
26. Rabinovich MI, Sokolov Y, Kozma R (2014) Robust sequential working memory recall in heterogeneous cognitive networks. Front Syst Neurosci 8:220
27. Tsuda I (2001) Toward an interpretation of dynamic neural activity in terms of chaotic dynamical systems. Behav Brain Sci 24:793–847
28. Haken H (1983) Synergetics: an introduction. Springer, Berlin
29. Bressler SL, Kelso JA (2001) Cortical coordination dynamics and cognition. Trends Cogn Sci 5(1):26–36
30. Kelso JAS, Engstrom DA (2006) The complementary nature. MIT Press, Cambridge
31. Tognoli E, Kelso JAS (2014) The metastable brain. Neuron 81(1):35–48
32. Freeman WJ (2003) The wave packet: an action potential for the 21st century. J Integr Neurosci 2:3–30
33. Kozma R, Puljic M, Balister P, Bollobas B, Freeman WJ (2005) Phase transitions in the neuropercolation model of neural populations with mixed local and non-local interactions. Biol Cybern 92:367–379
34. Werner G (2007) Metastability, criticality, and phase transitions in brains and its models. BioSystems 90(496–508):2007
35. Freeman WJ (2008) A pseudo-equilibrium thermodynamic model of information processing in nonlinear brain dynamics. Neural Netw 21:257–265
36. Freeman WJ, Holmes MD, West GA, Vanhatalo S (2006) Fine spatiotemporal structure of phase in human intracranial EEG. Clin Neurophysiol 117:1228–1243
37. Bak P (1996) How nature works the science of self-organized criticality. Springer, New York
38. Jensen HJ (1998) Self-organized criticality: emergent complex behavior in physical and biological systems. Cambridge University Press, New York
39. Beggs JM, Plenz D (2003) Neuronal avalanches in neocortical circuits. J Neurosci 23(35):11167–11177
40. Beggs JM (2008) The criticality hypothesis: how local cortical networks might optimize information processing. Philos Trans R Soc A: Math, Phys Eng Sci 366(1864):329–343
41. Haldeman C, Beggs JM (2005) Critical branching captures activity in living neural networks and maximizes the number of metastable states. Phys Rev Lett 94(5):058101
42. Sornette D, Quillon G (2012) Dragon-kings: mechanisms, statistical methods and empirical evidence. Eur Phys J Special Top 205(1):1–26
43. Puljic M, Kozma R (2005) Activation clustering in neural and social networks. Complexity 10(4):42–50
44. Erdi P, Kozma R, Puljic M, Szente J (2013) Neuropercolation and related models of criticalities. In: Contents XXIX-th European meeting of statisticians, Hungary, p 106

45. Kozma R, Puljic M (2015) Random graph theory and neuropercolation for modeling brain oscillations at criticality. Curr Opin Neurobiol 31:181–188
46. Zamora-Lopez G, Zhou C, Kurths J (2011) Exploring brain function from anatomical connectivity. Front Neurosci 5
47. Freeman WJ, Kozma R, Bollobas B, Riordan O (2009) Chapter 7. Scale-free cortical planar network. In: Bollobas B, Kozma R, Miklos D (eds) Handbook of large-scale random networks. Series: Bolyai mathematical studies, vol 18. Springer, New York, pp 277–324
48. Freeman WJ, Vitiello G (2010) Vortices in brain waves. Int J Mod Phys B 24(17):3269–3295
49. Freeman WJ, Burke BC, Holmes MD (2003) Aperiodic phase re-setting in scalp EEG of beta-gamma oscillations by state transitions at alpha-theta rates. Hum Brain Mapp 19(4):248–272
50. Kozma R, Freeman WJ (2008) Intermittent spatio-temporal de-synchronization and sequenced synchrony in ECoG signals. Chaos 18:037131
51. Freyer F, Aquino K, Robinson PA, Ritter P, Breakspear M (2009) Bistability and non-Gaussian fluctuations in spontaneous cortical activity. J Neurosci 29(26):8512–8524
52. Rice SO (1950) Mathematical analysis of random noise and appendixes. Technical Publications Monograph B-1589. Bell Telephone Labs Inc, New York
53. Freeman WJ (2009) Deep analysis of perception through dynamic structures that emerge in cortical activity from self-regulated noise. Cogn Neurodyn 3(1):105–116
54. Ruiz Y, Pockett S, Freeman WJ, Gonzales E, Guang Li (2010) A method to study global spatial patterns related to sensory perception in scalp EEG. J Neurosci Methods 191:110–118
55. Freeman WJ, Kozma R (2010) Freeman's mass action. Scholarpedia 5(1):8040

Chapter 4
Short and Long Edges in Random Graphs for Neuropil Modeling

4.1 Motivation of Using Random Graph Theory for Modeling Cortical Processes

Graph theoretical approaches have been very successful in the past decade to describe functional and structural aspects of brain networks; see, for example [1–5]. The existence of functional links between cortical nodes has been postulated in EEG, MEG, and fMRI data [6–11]. Small-world and scale-free networks have been studied extensively in real-world networks [12, 13] and they bear relevance to brain networks [14–16]. The presence of power-law structural properties in the cortex is a controversial issue [17, 18]. Experimental data are mostly counter-indicative of power-law distribution in connection lengths and degree sequences [19]. Some experiments indicate exponential length distribution of axons [20]. Small-world networks and highly connected hub structures, also called $RichClubs$, have been identified in a wide range of functional brain networks [21, 22].

Our goal is to describe structural and functional properties of the neuropil, the densely connected cortical tissue. We rely on random graph theory (RGT) developed in the past 50+ years [23–26]. Neuropil is the most complex substance in the known Universe and it contains massively interconnected neurons with excitatory and inhibitory links. A key task is modeling dynamic processes in brains with rapid conversion between high-dimensional and low-dimensional levels of organization. Sudden changes in spatio-temporal cortical dynamics resemble phase transitions in physical systems, but phase transitions in brains are much more complex than in physics [27]. In brains, highly organized metastable states are maintained for 100–200 ms, while the transitions from one state to another are rapid ($\sim$10 ms). Integro-differential equations are not suited well for modeling such discontinuous processes.

The theory of large-scale random graphs (RGT) and probabilistic cellular automata (PCA) provides an efficient approach to describe transitions between high- and low-dimensional spaces. Two contradictory aims are to be reconciled when developing

R. Kozma and W.J. Freeman, *Cognitive Phase Transitions in the Cerebral Cortex – Enhancing the Neuron Doctrine by Modeling Neural Fields*,
Studies in Systems, Decision and Control 39, DOI 10.1007/978-3-319-24406-8_4

mathematical models of complex systems, such as the cortex, using graph theory. The developed model should produce random graphs whose structure resembles that of the neocortex as much as possible, on the one hand, but it should not be so complicated that it is not susceptible to mathematical analysis, on the other hand. The hope is that the mathematical theory of random graphs will eventually be advanced enough that both requirements can be satisfied: a sophisticated model can be constructed that incorporates most of the requirements, and this very complicated model can be analyzed precisely, meaning that one can obtain mathematical descriptions both of its structure, and of the associated dynamics [14, 28]. RGT provides a solid foundation to describe the evolution of networks of continuously increasing size that can approach and exceed dynamic boundaries, thereby bridging the divide between the microscopic and macroscopic domains by summing rather than approximating. Therefore, RGT has the potential effectively describe the sudden changes and discontinuities observed in cortical processes by employing the powerful tools of random graphs and percolation theory.

When applying random graph theory to brains, cognitive processes are interpreted as intermittent percolation phenomena in the neuropil medium, and the corresponding approach is coined *neuropercolation* [14, 29, 30]. Importantly, neuropercolation relaxes some of the mathematical conditions defining percolation, therefore neuropercolation generalizes percolation to the broader class of probabilistic cellular automata. Mathematical results have been obtained on the onset of brief episodes of transient percolation processes progressing very rapidly through the PCA array after the emergence of some percolating sets, following a very long waiting period at some metastable states [31, 32].

Percolation transitions were studied in cultures of brain tissue and acute cortical slices [33]. Frantic and Milkovic [34] determined that spike packet propagation through synfire chains [35] exhibit critical behavior corresponding to percolation phase transitions. Theoretical work has been conducted on modeling criticality and phase transitions using concepts of percolation processes [36, 37]. RGT uses the rigorously defined concept of phase transition to describe intermittent dynamics in the neuropil medium.

4.2 Glossary of Random Graph Terminology

Graphs are defined as a collection of nodes (vertices) with links (edges, channels) among them. The term random refers to indeterminacy and the assignment of a probability value to the existence of a link between any pair of nodes, as well as to the probabilistic nature of the dynamic rules of the interaction between nodes. Early examples of random graphs are given by Rigorous work has been conducted on modeling criticality and phase transitions using concepts of percolation processes [23, 38]. We adopt Erdos-Renyi (ER) terminology:

- Graph: A mathematical object $G(n, M)$, with a collection of n vertices (nodes) and M edges (connections) between some pairs of the vertices.
- Directed graph (di-graph): A graph $G(n, M)$ with its edges directed from one node to the other.
- Random graph: Graph with its vertices and edges are selected some random way.
- Giant component: A component of the graph, which contains a constant (high) fraction of the nodes.
- Geometric graph: It has nodes placed in the n-dimensional Euclidean space.
- Path: A sequence of connecting edges between two nodes.
- Distance: The length of the shortest path between two nodes.
- Diameter: The length of the longest distance between any pairs of vertices of a graph.
- Degree of a node: The number of edges connected to that node. In directed graphs we distinguish in-degree and out-degree, i.e., the number of incoming and outgoing edges, respectively.
- Degree sequence: The non-decreasing sequence of the vertex degrees of the graph. The average degree is the average of the degree sequence.
- Scale-free network: A graph with a degree sequence distribution $d(k)$ following power law: $d(k)$ c/ka, where a is the exponent of the power law distribution.

ER random graphs have degree distributions converging to Poisson distribution with a dominant scale. Rigorous work has been conducted on modeling criticality and phase transitions using concepts of percolation processes [23]. The Erdos-Renyi random graphs are homogeneous, in the sense that the vertex set has no structure; in particular, all vertices are equivalent. Most networks in the real world are inhomogeneous, in the sense that different vertices may have very different properties. For example, in computer networks with hub structure, the hubs have much higher degree than other nodes. The vertex set has (usually geometric) structure, which influences the network structure: nearby vertices are more likely to be connected.

Many inhomogeneous random graph models have been introduced in the past decades. Rigorous work has been conducted on modeling criticality and phase transitions using concepts of percolation processes. Bollobás [39] provides a comprehensive review of research on inhomogeneous random graphs. That work gives a general formulation of sparse inhomogeneous random graphs, designed for representing graphs with geometric or similar structure. Here the term *sparse* means that the number of edges grows linearly with the number n of vertices, so the average degree is constant, but the definitions adapt immediately to other density ranges. In the fully dense range, with order n^2 edges. Note that the mathematical definitions of *sparse* and *dense* involve limits as n grows to infinity. The distinction may not be so clear in practice when modeling real-world graphs.

Real-life networks do not have a dominant scale, rather they often display power-law degree distribution. Preferential attachment is a widely used network modeling approach, when the probability of a new edge from a given node is proportional to the number of existing edges in that node. Barabasi and Albert [13] observed that growth with preferential attachment leads to scale-free degree distributions, with

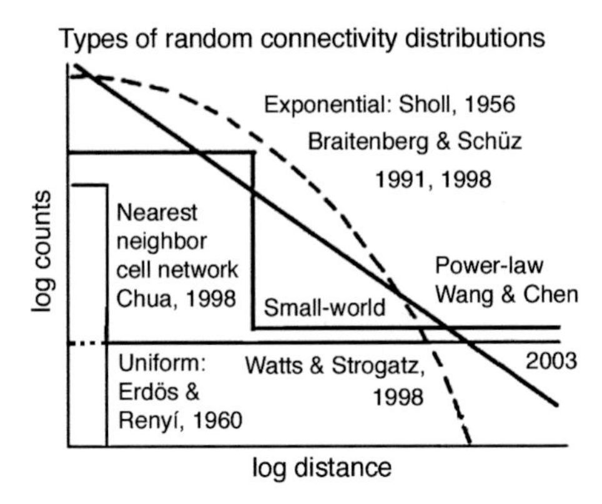

Fig. 4.1 Schematic distribution of the number of connections at varying distances in log–log coordinates. The *straight line* indicates the power-law distribution; uniform, nearest-neighbor, and small world distributions are illustrated as well; based on [51]

exponent $a \approx 2$. Under certain mathematical conditions, it was shown rigorously that $a = 2$ [40]; see also [41–44]. In addition, preferential depletion model has been found relevant to the models of neural systems [45].

Figure 4.1 illustrates some typical geometric graphs; ER random graphs have edge distribution that is independent of the geometry; see lower horizontal line in Fig. 4.1. Various alternative distributions are also displayed in Fig. 4.1. Two-dimensional cellular automata have a regular cellular lattice structure, when the connections are restricted to nearest or next nearest neighbor nodes (e.g., Chua 1998). Small-world (SW) networks have a combination of regular, short-rage connections, and a fraction of randomly selected non-local connections obtained by rewiring some edges [12]. Figure 4.1 also illustrates exponential distribution [20, 46] and scale-free length distribution, appearing as a linear curve in log–log display [13, 47–49].

With respect to inactivation of cortical areas transiently by electric or magnetic stimulation or permanently by lesions, Wang and Chen [48, 49] have commented on the relative immunity of scale-free networks to random loss of nodes compared with catastrophic disintegration of performance on loss of hubs. Extreme examples from clinical neurology are the syndromes resulting from loss of highly connected nodes in the midbrain reticular formation leading to coma, of the *substantia nigra* leading to Parkinson's disease, and the loss of the perforant path in Alzheimer's disease. More commonly the medical literature is rich with descriptions of loss of circumscribed cognitive functions, usually transitory, that accompany focal damage to limited areas of neocortex [50].

The view of neocortical connectivity as self-similar and scale-free (in the functional sense) suggests that every definable cognitive function has a hub in a location that is unique to the individual. Each hub is a kind of Achilles' heel in regard to loss of function upon focal damage, but the loss does not imply that the function is localized to the hub. On the contrary a hub operates only in the context of the macroscopic network. A lesion that does not remove a hub has little or no loss of function. The likelihood that each cognitive function has a different hub in a self-similar neocortical network might explain the variety of transient neurological syndromes associated with focal lesions in cortex.

4.3 Neuropercolation Basics

Deterministic cellular automata (CA): Let us consider a lattice in the d-dimensional Euclidean space, where the state of any lattice point is either active or inactive (1/0). In this review $d = 2$, i.e., we have a 2D layer of nodes modeling the layers of the cortex. The lattice is initialized with some configuration of active and inactive sites. The states of the lattice points are updated simultaneously using some deterministic rule that depends on the activations of their neighborhood. For example, majority voting is a popular choice of update rules, whereas a site's next state is given as the majority state of its neighbors. For related concepts, see cellular automata such as Conway's Game of Life, Chua's cellular neural networks, as well as thermodynamic models like the Ising model, and Hopfield nets [52–56].

Probabilistic Cellular Automata (PCA) are defined over similar configuration space as deterministic CA, but their update rule has a random component. Namely, each vertex is updated using a given rule based on its neighbors with probability $(1 - p)$ and takes its complements with probability p. Here p is typically a small value, which may be interpreted as a noise effect in brain tissues, describing random errors is the communication among the units and their inputs.

Bootstrap percolation is a mathematical approach studying activity propagation in graphs with a specific requirement on monotonicity of updates. Here we study bootstrap percolation over the same configuration space as in cellular automata with 0/1 activation levels. In the k-monotonous bootstrap percolation model, an active site always remains active, while an inactive site becomes active if at least k of its neighbors are active [57]. If the iterations ultimately lead to a configuration when active sites form cluster of infinite size, it is said that there is *percolation* in the lattice. Bootstrap percolation is concerned with the presence of percolation as the function of lattice dimension d, initial probability p, and neighborhood parameter k. It can be shown that on the infinite lattice Z^d, there exists a critical probability $p_c = f(d, k)$, that there is no percolation if $p \leq p_c$ and there is percolation with probability 1 if $p \geq p_c$. Generalizations of the original bootstrap percolation models are abundant. In a popular generalization approach active sites may flip to become inactive with probability p, which can model percolation in a polluted environment [58]. Accordingly, at every iteration step, an active site is removed with dilution probability q. I the case of the 2-dimensional lattice with the 2-neighbor rule ($k = 2$), the process percolates with probability one, if q/p^2 is small enough, and there is no percolation in the opposite case.

Phase Transition: The critical probability defines a phase transition between conditions leading to percolation and conditions which do not percolate. In the case of bootstrap percolation on the infinite lattice it can be shown that there exists a critical probability p_c, such that there is no percolation if $p < p_c$, and there is percolation with probability 1, if $p > p_c$. The critical probability defines a phase transition between conditions leading to percolation and conditions, which do not percolate [26]. For example, in the case of the 3-dimensional infinite lattice ($d = 3$), the local neighborhood consists of the 6 direct neighbors of the site, and itself. Selecting $k = 3$

means that an inactive site becomes active if at least 3 of its neighbors are active. It is shown that for $d = 3$, and $k = 3$ the critical probability is $p_c = 0$ [59].

In finite lattices: no clusters of infinite size may exist, and the above formal definition of phase transition does not apply. For the d-dimensional torus of size N Z_N^d, the probability of percolation is a continuous function of p. However, the probability of percolation rises rapidly from a value close to zero, to a value close to some function $p_c = f(N, d, k)$ when p increases [60]. Therefore, some threshold value can be introduced to define a transition, e.g., percolation probability of 0.5, or location of an inflexion point on the percolation probability curve, etc. In the case of some 3-dimensional finite lattices, Cerf and Cirillo [61] proved the conjecture [60], extending the above result of [59] that the threshold probability is of the order $1/\log\log N$, for a sequence of bootstrap percolation models as N grows to infinity. An example of percolation on the 2-dimensional torus, $d = 2$ and $k = 2$ is given in Fig. 4.2.

Neuropercolation describes generalizations of bootstrap percolation motivated by properties of the cortical tissue. Neuropercolation incorporates the following major conditions: (1) the presence of noise, (2) long axon effect, and (3) inhibition, as outlined below:

1. The dynamics of interacting neural populations is inherently nondeterministic due to dendritic noise and other random effects in the nervous tissue and external noise. This is expressed by Szentagothai "Whenever he is looking at any piece of neural tissue, the investigator becomes immediately confronted with the choice between two conflicting issues: the question of how intricate wiring of the neuropil is strictly predetermined by some genetically prescribed blueprint, and how much freedom is left to chance within some framework of statistical probabilities or some secondary mechanism of trial and error, or selecting connections according to necessities or the individual history of the animal" [62, 63]. Randomness plays a constructive role in neuropercolation and it can lead to the emergence of macroscopic order. The constructive role of noise resembles stochastic resonance (SR) [64]. An important difference from SR is the more intimate relationship between noise and the system dynamics in neuropercolation, due to the excitable nature of the neuropil [14, 65].
2. Neural populations stem ontogenetically in embryos from aggregates of neurons that grow axons and dendrites and form synaptic connections of steadily increasing density. In neural populations, most of the connections are short, but there are a relatively few long-range connections mediated by long axons [66]. The effect of long-range axons is related to small-world phenomena [12, 67].
3. The neural tissue contains two basic types of interactions: excitation and inhibition. Increased activity of excitatory units positively influences (excites) their neighbors, while highly active inhibitory units negatively influence (inhibit) the neurons they interact with. Mutual excitation is able to maintain stable background activity, while inhibition contributes to the emergence of sustained narrow-band oscillatory behavior in the neural tissue [68–70]. Inhibition in cortical

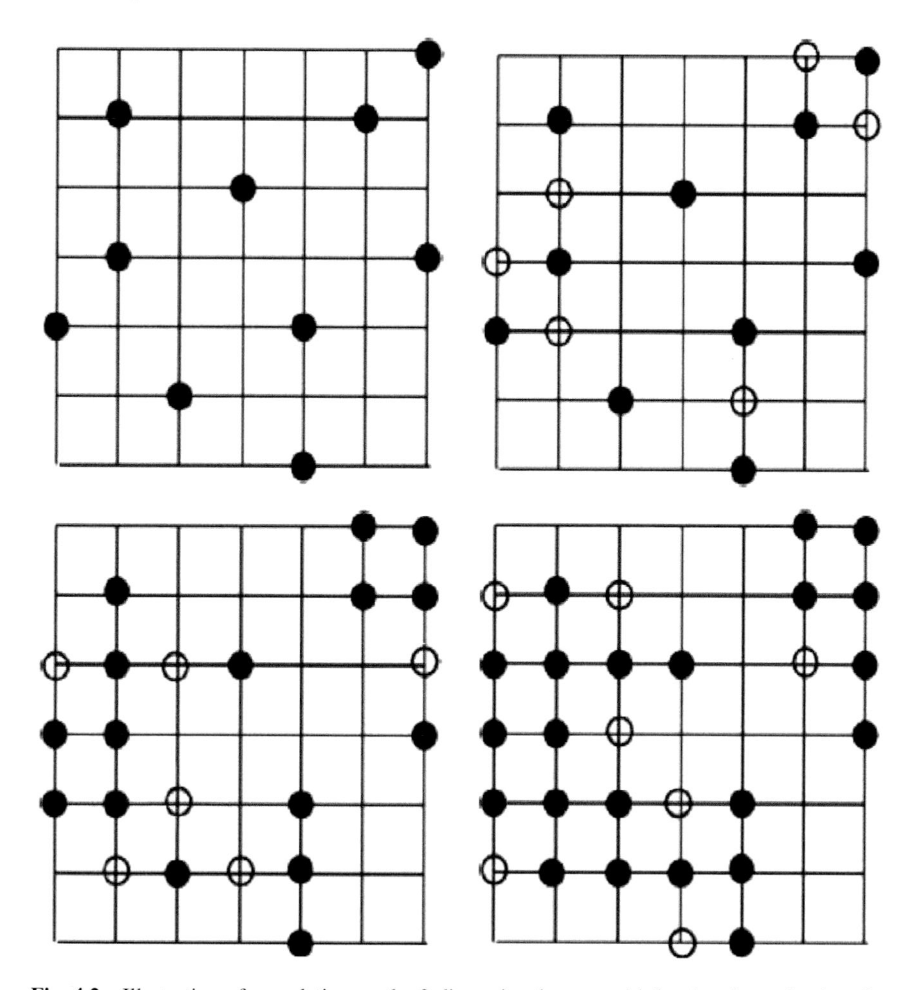

Fig. 4.2 Illustration of percolation on the 2-dimensional torus, with local update rule given by k = 2, i.e., a site becomes active (*black*) if at least 2 of its neighbors are active. The first 4 iteration steps are shown. At the 8th step all sites become active, i.e., the initial configuration percolates over the torus

tissues controls stability and metastability observed in brain behaviors [71–74]. Inhibitory effects are part of neuropercolation models.

Neuropercolation produces a system, which can maintain its state near criticality, and small changes in its internal and external state may induce conditions for rapid phase transitions from one dynamic state to the other. A key consequence of criticality is the potential manifestation of finite-size scaling theory and inherent coarse graining of the system characteristics scales, where the tools of renormalization group theory are applicable [75, 76]. Coarse graining reduces a highly complex system with many degrees of freedom to a sequence of smaller systems with smaller degrees of freedom by eliminating fluctuations at lower scales. An important property of phase transitions at the critical state is that the correlation length becomes infinite, or it becomes equal to the system size for finite systems. In the next discussions we summarize the mathematical and physical tools of critical analysis and finite size scaling theory. Further, we explore the hypothesis that EEG/ECoG results manifesting sudden changes in synchronization and rapid propagation of signals and phase dispersion across large-scale cortical areas are in fact manifestations of criticality and natural coarse graining in the brain as a large-scale cortical network graph. In this approach, the level of noise, rewiring, and strength of inhibition are parameters, which would control the dynamics of brains as they intermittently approach and leave critical states in their metastable regime of operation.

4.4 Critical Behavior in Neuropercolation with Mean-Field, Local, and Mixed Models

In a lattice with n nodes, activation density is defined as $\rho = 1/n \sum_{i=1}^{n} a_i(t)$, where $a_i(t)$ is the binary $(0/1)$ activation value of the ith node at time t. The density acts as an order parameter and it can exhibit various dynamic behaviors depending on the probabilistic update rules and the type of neighborhoods. There are geometric random graph models based on local, mean-field, and mixed neighborhoods.

- In local (nearest neighbor) models, each node has a neighborhood defined as those nodes that immediately follow/precede that node in any direction. For example in a 2-dimensional lattice, a node has 4 local neighbors, plus a loop connection to itself.
- In mean field models, the activations of randomly selected nodes are used in the update rule, instead of specified local neighborhoods. Since there is no ordering of the neighbors in mean-field approach, the updates depend only on the number of active states in the selected neighborhood.
- To generate mixed neighborhoods, we start with local neighbors then we substitute (rewire) some of the connections by connections selected randomly from the whole lattice.

4.4.1 Mean-Field Approximation

First we address the mean-field (MF) version of the majority PCA. MF models are not directly applicable to actual neural systems. However, MF models have been studied rigorously and it is anticipated that some of the conceptual results carry over to PCA and neuropercolation. In the mean field model, instead of considering the number of active nodes (k) in the specified neighborhood, the activations of randomly selected grid nodes are used in the update rule. Since there is no ordering of the neighbors, the transition probabilities depend only on the number of active states in the selected k-tuples. It is clear that the mean field model does not depend on the topology of the grid.

Considering a 2D torus of size NxN, the density of active points is defined as $d(t) = A(t)/N^2$, where $A(t)$ is the number of active nodes on the torus at time t. The density acts as an order parameter and it can exhibit various dynamic behaviors depending on the probabilistic rules, including the noise component p of the update rule. Mean field models have at least one stable fixed point and can have several stable and unstable fixed points, limit cycles, and chaotic oscillations. In mean field models with noise, various conditions have been derived for stable fixed-point solutions, and the presence of phase transitions between stable fixed points have been proven in mean field models with noise. In the case of the isotropic mean-field model, there is one fixed point at $\rho(t) = 0.5$ for noise (p) exceeding critical level $(p_c > p)$. For subcritical noise $(p < p_c)$, the fixed point at $\rho(t) = 0.5$ is unstable, and there are two other fixed points that are stable [31]. In the 2-dimensional lattice with majority update rule, the critical probability is $p_c = 7/30$. Recent progress in mean field models includes the rigorous proof of the existence of phase transition in the presence of a mix of short and long connections [77]. As an example, we show the analytical solution of critical behavior in MF model without long connections. Indeed, the condition for the fixed point density is given as [14]:

$$d = (1-\varepsilon)\left(\sum_{i=0}^{\lfloor|\Lambda|/2\rfloor} B(|\Lambda|, i) d^{|\Lambda|-i}(1-d)^i\right) + \varepsilon\left(1 - \sum_{i=0}^{\lfloor|\Lambda|/2\rfloor} B(|\Lambda|, i) d^{|\Lambda|-i}(1-d)^i\right),$$

where $B(a, b)$ is the binomial coefficient. Considering the cardinality of the neighborhood $|\Lambda| = k = 5$, the condition for stable solution is obtained as $p_c = 0.5 - 1/(12d^4 - 24d^3 + 8d^2 + 4d + 2)$. Substituting the value $d = 0.5$, we get $p_c = 7/30$. The above equation describing the fixed points approximates power law relationship with very good accuracy as $p \to p_c$:

$$|d - 0.5| \propto (\varepsilon_c - \varepsilon)^\beta, \tag{4.1}$$

where $\beta \approx 0.5$. This result says that near the critical point, the density as an order parameter has a power low relationship with $(p - p_c)$, where the power exponent β is $1/2$. This is an example of the universal critical behavior, and it has important consequences to more general RGT models as well.

Figure 4.3 illustrates the stable density values for $0 \le \varepsilon \le \varepsilon_c$ as described by the above equation.

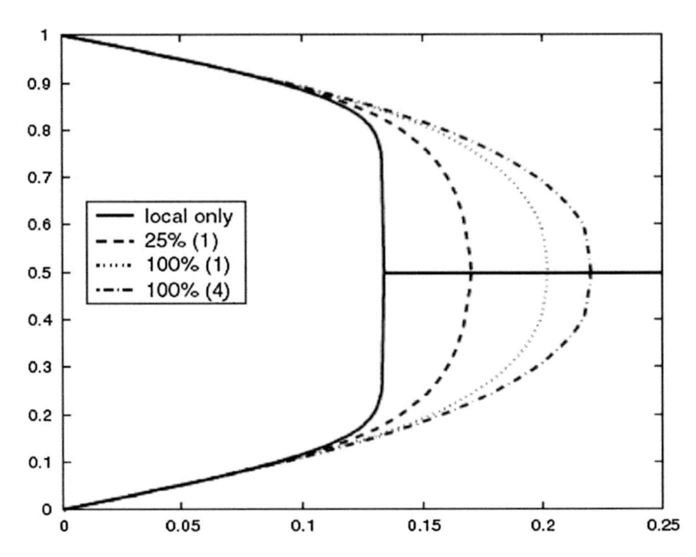

Fig. 4.3 Critical behavior of random bootstrap percolation for 2D homogeneous population. (a) Exact theoretical curves in the case of mean-field model. There is a phase transition at $\varepsilon_c = 7/30$. The model exhibits standard scaling properties near the critical point, as known in Ising models (Balister et al., 2006). (b) Neuropercolation simulation with various levels of rewiring in a 2-dimensional lattice; solid line: local only (no rewiring); *dash-dot line*: 100 % rewiring (mean field model). *Dashed lines* indicate various intermediate configuration with partial rewiring

4.4.2 Mixed Short and Long Connections

Rigorous mathematical analysis of phase transitions in (deterministic or probabilistic) cellular automata is very difficult, and only very few exact results are known. The existence of phase transitions has been proven for a few specific deterministic cellular automata [78, 79]. In majority PCA models with very small noise ($p \ll 1$), it has been shown that the model spends a long time in either low- or high-density configurations, before the very rapid transition to the other state [32]. The duration of the metastable states is exponentially long in terms of the system size, while the rapid transition between states happens much faster, i.e., in polynomial time. Majority PCA is much more general mathematical object than percolation, however, the rapid transition between metastable states can be approximated as bootstrap percolation starting from certain well-defined configurations (percolating sets) and propagating through the lattice [44]. Transitions for a single channel and for the ensemble average over a lattice without rewiring are shown in Fig. 4.4a. Results with edge rewiring are illustrated in Fig. 4.4b.

Mixed models pose very difficult mathematical problems and no exact mathematical proof of phase transitions exist in such systems. In the absence of analytical solutions, Monte Carlo simulations are used. An example of critical transitions is given in Fig. 4.5 using PCA simulation over a homogeneous layer of excitatory

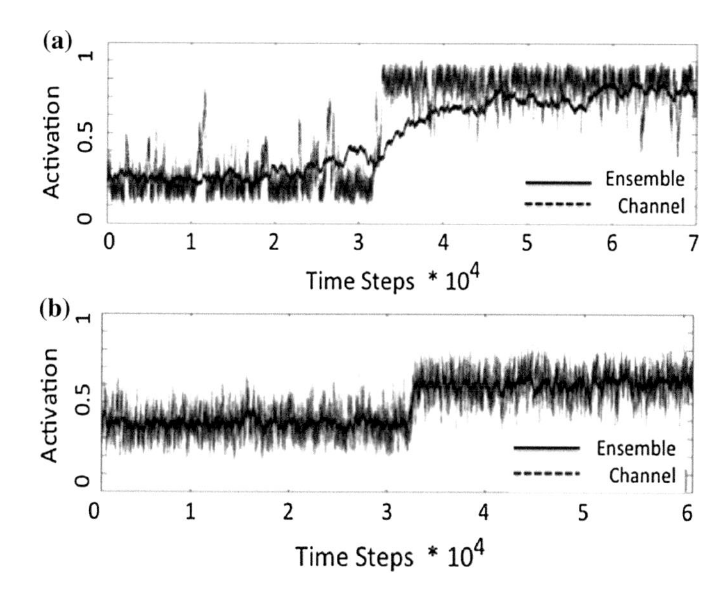

Fig. 4.4 Typical dynamics with rapid transitions in the two-dimensional lattice without rewiring **a** and with-rewiring **b**; periodic boundary condition applies. The sudden change in the mean activation may suggest that the observed effect is phase transition, although no rigorous mathematical proof exists at present

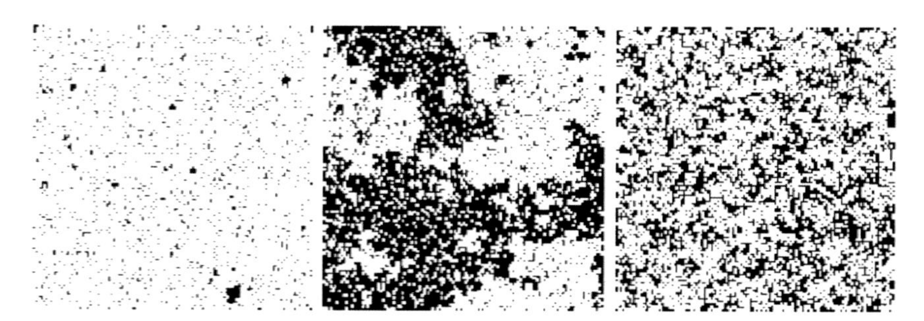

Fig. 4.5 Illustration of transitions in a PCA over a 2-dimensional lattice as a function of the noise level (p); *black* and *white dots* illustrate active and inactive sites, respectively. Simulations are with probabilistic majority rule without rewiring. The lattice size is 128×128. The leftmost frame shows a state with fragmented components (subcritical regime). The rightmost plate shows a population without spatial structure (supercritical regime). The central frame is an example of the critical state, when a huge cluster (giant component) emerges across the lattice [14]

components. After identifying critical parameters, the system behavior can be tuned and rapid transitions can be induced [28, 80–83]. There is a giant cluster spreading across the lattice in the critical case, central plot in Fig. 4.5. The components at criticality are expected to show fractal patterns with fractal dimension characterizing the universality class [76]. For details of the clustering aspects of PCA models, see [81, 84].

4.5 Finite Size Scaling Theory of Criticality in Brain Models

We adopt methods of statistical physics, which have been developed earlier for Ising spin glass systems [85]. Here we summarize the main ideas leading to the concept of neuropercolation and phase transition in the cortex.

It is of interest to evaluate lattice dynamics with noise p. If the number of active and inactive sites equals, the activation density satisfies $\rho(t) = 0.5$. This corresponds to a basal state in magnetic materials with no magnetization. Deviations from the 0.5 level gives the magnitude of the magnetization as: $m(t) = |\rho(t) - 0.5|$. Magnetization acts as an order parameter in our system. The expected value of magnetization is denoted as $\langle m \rangle$. In addition to magnetization, important quantities analyzed in Binder's method include the correlation length and susceptibility. The correlation length corresponds to the typical cluster size. The susceptibility describes the variance of the system activation per site. For Ising systems, magnetization, susceptibility, and correlation length satisfy a power law scaling behavior near criticality. Details of the corresponding statistical method of finite size scaling for the RGT model are given in the supplementary section.

References

1. Eguiluz VM, Chialvo DR, Cecchi GA, Baliki M, Apkarian AV (2005) Scale-free brain Functional networks. Phys Rev Lett 94(1):018102
2. Reijneveld JC, Ponten SC, Berendse HW, Stam CJ (2007) The application of graph theoretical analysis to complex networks in the brain. Clin Neurophysiol 118(11):2317–2331
3. Turova TS, Villa AE (2007) On a phase diagram for random neural networks with embedded spike timing dependent plasticity. Biosystems 89(1):280–286
4. Tlusty T, Eckmann JP (2009) Remarks on bootstrap percolation in metric networks. J Phys A Math Theor 42(20):205004
5. Gallos LK, Makse HA, Sigman M (2012) A small world of weak ties provides optimal global integration of self-similar modules in functional brain networks. Proc Natl Acad Sci 109(8):2825–2830
6. Sporns O, Tononi G, Kotter R (2005) The human connectome: a structural description of the human brain. PLOS Comput Biol 1(4):245–251
7. Honey CJ, Thivierge JP, Sporns O (2010) Can structure predict function in the human brain? Neuroimage 52(3):766–776
8. Stam CJ, Jones BF, Nolte G, Breakspear M, Scheltens P (2007) Small-world networks and functional connectivity in Alzheimer's disease. Cereb Cortex 17:92–99
9. Bonifazi P, Goldin M, Picardo MA, Jorquera I, Cattani A, Bianconi G, Cossart R (2009) GABAergic hub neurons orchestrate synchrony in developing hippocampal networks. Science 326(5958):1419–1424
10. Deco G, Corbetta M (2011) The dynamical balance of the brain at rest. Neuroscientist 17(1):107–123
11. Kim DJ, Bolbecker AR, Howell J, Rass O, Sporns O, Hetrick WP, O'Donnell BF (2013) Disturbed resting state EEG synchronization in bipolar disorder: a graph-theoretic analysis. NeuroImage Clin 2:414–423
12. Watts DJ, Strogatz SH (1998) Collective dynamics of small-world networks. Nature 393:440–442

13. Barabasi A-L, Albert R (1999) Emergence of scaling in random networks. Science 286:509–512
14. Kozma R, Puljic M, Balister P, Bollobas B, Freeman WJ (2005) Phase transitions in the neuropercolation model of neural populations with mixed local and non-local interactions. Biol Cybern 92:367–379
15. Freeman WJ, Kozma R, Bollobas B, Riordan O (2009) Chapter 7. Scale-free cortical planar network. In: Bollobas B, Kozma R, Miklos D (eds) Handbook of large-scale random networks, vol 18., Series: bolyai mathematical studiesSpringer, New York, pp 277–324
16. Kello CT, Brown GD, Ferrer-i-Cancho R, Holden JG, Linkenkaer-Hansen K, Rhodes T, Van Orden GC (2010) Scaling laws in cognitive sciences. Trends Cogn Sci 14(5):223–232
17. Breakspear M (2004) Dynamic connectivity in neural systems: theoretical and empirical considerations. Neuroinformatics 2(2):205–225
18. Chen Q, Shi D (2004) The modeling of scale-free networks. Phys A 333:240–248
19. Sporns O, Chialvo DR, Kaiser M et al (2004) Organization, development, and function of complex brain networks. Trends Cogn Sci 8(9):418–425
20. Braitenberg V, Schuz (1998) Cortex: statistics and geometry of neuronal connectivity, 2nd edn. Springer, Berlin
21. Zamora-Lopez G (2009) Linking structure and function of complex cortical networks. Ph.D. thesis, University of Potsdam, Potsdam
22. Bullmore E, Sporns O (2012) The economy of brain network organization. Nat Rev Neurosci 13(5):336–349
23. Erdos P, Renyi A (1959) On random graphs. Publ Math Deb 6:290–297
24. Erdos P, Renyi A (1960) On the evolution of random graphs. Publ Math Inst Hung Acad Sci 5:17–61
25. Bollobas B (1985/2001) Random graphs. Cambridge studies in advanced mathematics. Cambridge University Press, Cambridge
26. Bollobas B, Riordan O (2006) Percolation. Cambridge University Press, Cambridge
27. Plenz D, Thiagarajan TC (2007) The organizing principles of neuronal avalanches: cell assemblies in the cortex? Trends Neurosci 30:101110
28. Kozma R, Puljic M (2015) Random graph theory and neuropercolation for modeling brain oscillations at criticality. Curr Opin Neurobiol 31:181–188
29. Kozma R, Puljic M, Balister P, Bollobas B, Freeman WJ (2004) Neuropercolation: a random cellular automata approach to spatio-temporal neurodynamics. Lect Notes Comput Sci LNCS 3305:435–443
30. Kozma R (2007) Neuropercolation. Scholarpedia 2(8):1360
31. Balister P, Bollobas B, Kozma R (2006) Large-scale deviations in probabilistic cellular automata. Random Struct Algorithm 29:399–415
32. Balister P, Bollobas B, Johnson JR, Walters M (2010) Random majority percolation. Random Struct Algorithm 36(3):315–340
33. Breskin I, Soriano J, Moses E, Tlusty T (2006) Percolation in living neural networks. Phys Rev Lett 97(18):188102
34. Franovic I, Milkovic V (2009) Percolation transition at growing spatiotemporal fractal patterns in models of mesoscopic neural networks. Phys Rev E 79(6):061923
35. Gewaltig MO, Diesmann M, Aertsen A (2001) Propagation of cortical synfire activity: survival probability in single trials and stability in the mean. Neural Netw 14(6):657–673
36. Turova TS (2012) The emergence of connectivity in neuronal networks: from bootstrap percolation to auto-associative memory. Brain Res 1434:277–284
37. Turova T, Vallier T (2015) Bootstrap percolation on a graph with random and local connections. Preprint http://arxiv.org/abs/1502.0149arXiv:1502.01490
38. Gilbert EN (1959) Random graphs. Ann Math Stat 30(4):1141–1144
39. Bollobás B, Janson S, Riordan O (2007) The phase transition in inhomogeneous random graphs. Random Struct Algorithm 31(1):3–122
40. Bollobas B, Riordan O (2003) Results on scale-free random graphs. Handbook of graphs and networks, Wiley, Weinhiem, pp 1–34

41. Albert R (2002) Statistical mechanics of complex networks. Rev Mod Phys 74:47–97
42. Dorogovtsev SN (2003) Mendes JFF (2003) Evolution of networks: from biological nets to the Internet and WWW. Oxford University Press, Oxford
43. Newman M, Barabasi A-L, Watts DJ (eds) (2006) The structure and dynamics of networks. Princeton Studies in Complexity, Princeton University Press, Princeton, p x+582
44. Bollobas B, Kozma R, Miklos D (eds) (2009) Handbook of large-scale random networks. Bolyai society mathematical studies. Springer, New York
45. Schneider CM, De Arcangelis L, Herrmann HJ (2011) Scale-free networks by preferential depletion. Europhys Lett. 95(1):16005
46. Sholl DA (1956) The organization of the cerebral cortex. Methuen-Wiley, London
47. Barabasi A-L (2002) Linked. The new science of networks. Perseus, Cambridge
48. Wang XF, Chen G (2002) Synchronization in scale-free dynamical networks: robustness and fragility. IEEE Trans Circuits Syst Fundam Theory Appl 49:54–62
49. Wang XF, Chen GR (2003) Complex networks: small-world, scale-free and beyond. IEEE Trans Circuits Syst 31:6–20
50. Critchley EM (1979) Drug-induced neurological disease. BMJ 1(6167):862–865
51. Freeman WJ, Breakspear M (2007) Scale-free neocortical dynamics. Scholarpedia 2(2):1357
52. Berlekamp, ER, Conway JH, Guy RK (1982) Winning ways for your mathematical plays, volume 1: games in general, Academic Press, New York
53. Hopfield JJ (1982) Neuronal networks and physical systems with emergent collective computational abilities. Proc Natl Acad Sci USA 81:3058–3092
54. Kauffman S (1993) The origins of order—self-organization and selection in evolution. Oxford University Press, Oxford
55. Chua LO (1998) CNN. A paradigm for complexity. World Scientific, Singapore
56. Wolfram S (2002) A new kind of science. Wolfram Media Inc., Champaign
57. J Phys A (1988) Metastability effects in bootstrap percolation. 21:3801–3813
58. Gravner J, McDonald E (1997) Bootstrap percolation in a polluted environment. J Stat Phys 87(3–4):915–927
59. Schonmann R (1992) On the behavior of some cellular automata related to bootstrap percolation. Ann Probab 20(1):174–193
60. Adler J (1991) Bootstrap percolation. Phys A 171:453–470
61. Cerf R, Cirillo EN (1999) Finite size scaling in three-dimensional bootstrap percolation. Ann Probab 27(4):1837–1850
62. Szentagothai J (1978) Specificity versus (quasi-) randomness in cortical connectivity. In: Brazier MAB, Petsche H (eds) Architectonics of the cerebral cortex connectivity. Raven Press, New York, pp 77–97
63. Szentagothai J (1990) Specificity versus (quasi-) randomness revisited. Acta Morphol Hung 38:159–167
64. Bulsara A, Gammaitoni L (1996) Tuning in to noise. Phys Today 49(3):39–45
65. Kozma R (2003) On the constructive role of noise in stabilizing itinerant trajectories on chaotic dynamical systems. Chaos 11(3):1078–1090
66. Das A, Gilbert CD (1995) Long-range horizontal connections and their role in cortical reorganization revealed by optical recording of cat primary visual cortex. Nature 375(6534):780–784
67. Strogatz SH (2001) Exploring complex networks. Nature 410(6825):268–276
68. Aradi I, Barna G, Erdi P (1995) Chaos and learning in the olfactory bulb. Int J Intell Syst 10:89
69. Freund TF, Buzsáki G (1996) Interneurons of the hippcampus. Hippocampus 6:347–470
70. Arbib M, Erdi P, Szentagothai J (1997) Neural organization. MIT Press, Cambridge
71. Kelso JAS (1995) Dynamic patterns: the self organization of brain and behavior. MIT Press, Cambridge
72. Xu D, Principe JC (2004) Dynamical analysis of neural oscillators in an olfactory cortex model. IEEE Trans Neural Netw 15(5):1053–1062
73. Ilin R, Kozma R (2006) Stability of coupled excitatory-inhibitory neural populations application to control multistable systems. Phys Lett A 360:66–83
74. Kelso JAS, Engstrom DA (2006) The complementary nature. MIT Press, Cambridge

75. Kadanoff LP, Ceva H (1971) Determination of an operator algebra for a two-dimensional ising model. Phys Rev B 3:3918
76. Odor G (2004) Universality classes in nonequilibrium lattice systems. Rev Mod Phys 76:663–724
77. Janson S, Kozma R, Ruszinkó M, Sokolov Y (2015) Activation process on a long-range percolation graph with power law long edge distribution. Part I phase transition without inhibition (in progress)
78. Toom AL (1980) Stable and attractive trajectories in multicomponent systems. Multicomponent Random Syst Adv Probab Relat Top 6:540–575
79. Gacs P (2001) Reliable cellular automata with self-organization. J Stat Phys 103:45–267
80. Puljic M, Kozma R (2008) Narrow-band oscillations in probabilistic cellular automata. Phys Rev E 78:026214
81. Puljic M, Kozma R (2010) Broad-band oscillations by probabilistic cellular automata. J Cell Autom 5(6):491–507
82. Kozma R, Puljic M (2013) Learning effects in neural oscillators. Cogn Comput 5(2):164–169
83. Kozma R, Puljic M (2013) Hierarchical random cellular neural networks for system-level brain-like signal processing. Neur Netw 45:101–110
84. Puljic M, Kozma R (2005) Activation clustering in neural and social networks. Complexity 10(4):42–50
85. Binder K (1981) Finite scale scaling analysis of ising model block distribution function. Z Phys B 43:119–140

Chapter 5
Critical Behavior in Hierarchical Neuropercolation Models of Cognition

5.1 Basic Principles of Hierarchical Brain Models

Here we introduce a hierarchical approach to brain dynamics using Freeman K sets, including the hierarchy of $K0$, *KI*, *KII*, and *KIII* sets [1]. They correspond to brain scales starting from the sub-mm range to the complete hemisphere, for details see supplementary sections.

At the base of the K hierarchy are $K0$ populations of non-interacting neurons. $K0$ units form cortical micro-columns of about 10^4 neurons. $K0$ nodes model dendritic integration in average neurons and a sigmoid static nonlinearity for axon transmission. $K0$ sets are governed by point attractors with zero output and stay at equilibrium except when perturbed. An interactive *KI* set represents a 2-dimensional layer of $K0$ sets, which may consist of either excitatory (KI_e) or inhibitory (KI_i) units. A major function of the inhibitory neural populations is to selectively enhance sensitivity to reinforced conditional stimuli (CS) under pair-wise excitation. If a KI_e set has sufficient connection density, it maintains a non-zero state of background activity by mutual excitation.

A *KII* set represents a collection of excitatory and inhibitory populations, KI_e and KI_i. Examples include the olfactory bulb, hippocampal regions, the prepyriform cortex, and various areas of the neocortex. Due to the negative feedback between neural populations, *KII* sets may undergo a transition from a point attractor to narrow-band oscillations in the beta-gamma range. The *KIII* set consists of several interconnected *KII* sets in a multi-layer structure, and it describes various sensory system in brains, e.g., olfactory, visual, auditory, and somatosensory modality. *KIII* sets generate broadband, aperiodic oscillations as background activity by combined negative and positive feedback among several *KII* populations with incommensurate frequencies. Sensory systems create spatial patterns of macroscopic neural activity from the microscopic information that was initiated in sensory receptors, and that was preprocessed in the olfactory bulb for smell and the thalamus for all other senses. Negative feedback loops provide the narrow-band carrier wave that transports the

R. Kozma and W.J. Freeman, *Cognitive Phase Transitions in the Cerebral Cortex – Enhancing the Neuron Doctrine by Modeling Neural Fields*,
Studies in Systems, Decision and Control 39, DOI 10.1007/978-3-319-24406-8_5

cortical output and parses it into frames [2]. The increase in nonlinear feedback gain results in the destabilization of the background activity and leads to emergence of a spatial amplitude modulation (AM) patterns. Positive feedback loops provide the local amplification and attenuation of amplitude of the carrier wave, which results in the spatial texture of AM patterns and expresses the memory of stimuli that is stored in the modifications of synapses among excitatory cells. The inhibitory positive feedback loops provide the lateral inhibition that sharpens the spatial AM and facilitates contrast enhancement and the emergence of Mach band patterns.

5.2 Narrow-Band Oscillations in Lattices with Inhibitory Feedback

In a more realistic model of the cortical tissue, inhibitory populations are introduced, in addition to excitatory units. Inhibitory effects lead to the emergence of sustained narrow-band oscillations in the neural tissue, which are preconditions of more complex, multicomponent brain rhythms. In local and mixed models with inhibition, rich dynamics with stable and unstable fixed points and narrow-band oscillations take place [3].

In local and mixed models with inhibition, lattice simulations provide insight to the spatio-temporal dynamics. Detailed numerical evaluations show the occurrence of additional stable and unstable fixed points, leading to multi-stability and to the onset of narrow-band oscillations. When the noise level is varied while all other parameters are fixed, two critical points have been identified, which demarcate the parameter region with narrow-band oscillations. The first critical point marks the onset of prominent narrow-band oscillations, while the second critical point describes the transition point where narrow-band oscillations disappear [4]. In the case with inhibition, we have an extended critical region characterized by narrow-band oscillations. For subcritical regimes, multi-modal oscillations dominate the dynamics.

Example of Monte Carlo experiments with *KII* exhibiting damped oscillations is given in Fig. 5.1. When the noise level is varied, while all other parameters are fixed in the presence of inhibition, two critical points have been found, 0 and 1 [4]. These critical points demarcate the region, in which large-scale narrow-band oscillations exist. 0 marks the transition from unimodal regime to bimodal (limit cycle) oscillations, and 1 indicates the transition from bimodal to quadro-modal distribution where large-scale synchronized oscillations diminish. The emergence of a region of large-scale synchrony with prominent narrow-band oscillations is crucial for building a hierarchy of cognitively relevant brain models using *KII* populations.

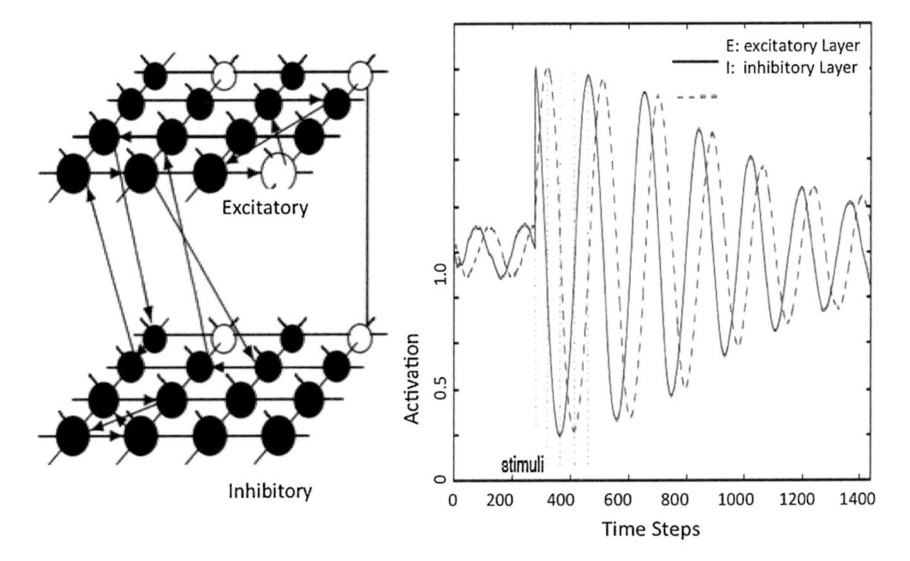

Fig. 5.1 Example of (KI)I oscillators in neuropercolation. *Left panel* illustration of rewiring a few edges between the excitatory and inhibitory layers. *Right panel* damped periodic oscillations are seen after impulse perturbation at criticality. There is a quarter cycle delay between the excitatory and inhibitory dynamics [4]

5.3 Broad-Band Oscillations in Coupled Multiple Excitatory-Inhibitory Layers

Consider the *KIII* model of three coupled oscillators; see Fig. 5.2. This system exhibits four possible critical points, ε_0, ε_1, ε_2, and ε_3 [5]. The case $\varepsilon_0 < \varepsilon < \varepsilon_1$ is of special interest to neural modeling, when the oscillators produce large-scale synchronization with narrow-band, bimodal oscillations. Outside the region of large-scale synchronization, there is either broad-band background activity ($\varepsilon_1 < \varepsilon < \varepsilon_2$), see Fig. 5.1, bottom time series for Oscillator 3; or the mix of narrow-band and broad-band oscillations ($\varepsilon_2 < \varepsilon < \varepsilon_3$), or unimodal regime ($\varepsilon < \varepsilon_0$).

It has been shown that Hebbian learning can shift the operational mode into and out-of the regime of large-scale synchronized oscillations $\varepsilon_0 < \varepsilon < \varepsilon_1$. This result is an important manifestation of the formation of Hebbian assemblies and their effect in facilitating the transition between broad-band background activity and narrow-band (limit cycle) response regimes [6].

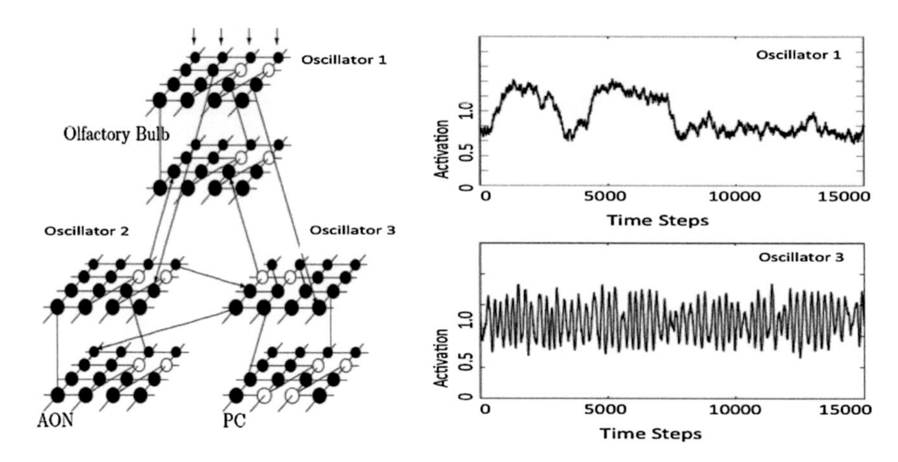

Fig. 5.2 Illustration of the 6-layer (KI)II structure of the olfactory system. There are excitatory-inhibitory pairs of layers in the Olfactory Bulb (OB), the Anterior Olfactory Nucleus (AON), and the Prepyriform Cortex (PC). The right panel shows the generated oscillations in the excitatory layers of Oscillator 1 and 3, respectively; from [6]

5.4 Exponentially Expanding Graph Model

Geometric graphs describe networks embedded in physical space. They allow studying edge length distributions, in addition to degree distributions. The cortical tissue can be approximated as a 2-dimensional sheet with multi-layer architecture, which can be modeled by 2D geometric graphs. To model spatial development of brains from the inception of the individual to the formation of detailed brain structures, we evoke the concept of *pioneer neurons* [7, 8] and the transitory subplate [9, 10]. Exponential distributions have been reported in the lengths of axon collaterals [11] and projections [12], which display predominantly local connections and increasingly sparse connections with increasing distance from the cell bodies. The formation of input and output pathways of cortical sensory and motor areas is dominated by topographic mapping, which arises by virtue of the manner in which pioneer neurons form initial connections that are followed thereafter by successions of new arrivals. The outcome is the formation of nodes that are not neurons but are local populations of neurons, manifested in dendritic bundles, barrels, patches, hypercolumns, and similar cyto-architectonic configurations.

We have developed the *Exponentially Expanding Graph model* for brains, based on principles of planar geometry [13], which does not require the assumption of preferential attachment but it produces robust scale-free connection length distribution. The development of brain structures is described as the evolution that starts from an initial set of N pioneer neurons. N is typically a small number compared to the size of the fully developed human brain (10^{11} neurons). Pioneer neurons are modeled as small balls densely packed in a small area of characteristic size; see G_0 in Fig. 5.3. As the cortex develops, the space spanned by the neurons expands so that the

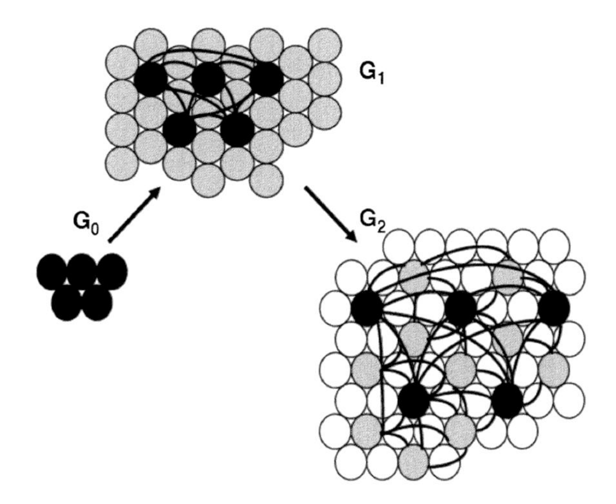

Fig. 5.3 Illustration of the sequential evolution of the exponentially expanding graph model; *black circles*—pioneer neurons G_0, *grey* and *white circles*—consecutive generations of neurons. The first two steps of the evolution are shown with G_1 and G_2. The illustrated regular structure is used only for simplicity [14]

distance between the original neurons increases. The space created by this expansion is filled with the newly created neurons. These new neurons grow connections to the existing ones. In the exponential model, an arriving new node makes connection to the existing neurons in its neighborhood within a disk D. A schematic view of the initial stage of development, from G_0, G_1, to G_2 graphs is shown in Fig. 5.3.

It is shown mathematically that the in-degree distribution follows exponential distribution, i.e., the number of vertices with in-degree at least k is proportional to $\exp(-2k/D)$ [13]. The edge length distribution follows a power law: the fraction of edges of length at least r is proportional to $1/r^2$. This power law is valid from the initial neighborhood given by disk D, to the macroscopic size of the population. The exponent is 2, which equals to the dimension of the space in which the evolving neuronal population lives.

The presence of scale-free structure of connection lengths and degree distribution is an open question in the brain network literature. Robust signal transmission can benefit from anatomical structures reflecting certain scale-free connectivity, and it is just natural to explore if brains exhibit such behavior. A key advantage of scale-free networks is their robustness to random failure of components. However, scale-free structure can become a liability if a mechanism systematically targets and damages hubs. In order to provide resilience to loss of hubs, some overlap of the hubs may be beneficial. The exponentially expanding graph model provides a solution to this question through the key role of pioneer neurons in the cortical tissue. They remain connected even as they spread from each other spatially during the brain network exponential expansion; see in Fig. 5.3 the highly connected pioneer neurons (black circles) of graph G_2. Introduced into this scale-free manifold are Hebbian nerve cell assemblies, which serve as hubs of tightly interconnected neurons, and which by mutual excitation enact generalization and abstraction over categories of equivalent input neurons [15]. The special connectivity structure due to pioneer neurons may be dormant in non-critical situations, but it may become dominant in the case of sudden changes or external disturbances when rapid response is the key.

Rich Club (RC) networks [16] describe a related concept. RC is a network property that takes place when the network hubs, i.e., the nodes with the largest number of neighbors, are densely interconnected [17]. In this sense, the exponentially expanding graph model model manifests RC property due to the special role of pioneer neurons. In addition to the RC feature, the exponentially exploding brain model reflects important experimentally observed properties of brain networks, including short processing paths, the existence of massive parallel processing paths, and the emergence of hub structures with modular architecture. Ongoing studies are expected to enhance recent results in brain connectome studies [18, 19].

References

1. Freeman WJ, Erwin H (2008) Freeman K-set. Scholarpedia 3(2):3238
2. Freeman WJ (2005) Origin, structure, and role of background EEG activity. Part 3. Neural frame classification. Clin Neurophysiol 116(5):1118–1129
3. Ilin R, Kozma R (2006) Stability of coupled excitatory-inhibitory neural populations application to control multistable systems. Phys Lett A 360:66–83
4. Puljic M, Kozma R (2008) Narrow-band oscillations in probabilistic cellular automata. Phys Rev E 78:026214
5. Puljic M, Kozma R (2010) Broad-band oscillations by probabilistic cellular automata. J Cell Autom 5(6):491–507
6. Kozma R, Puljic M (2013) Learning effects in neural oscillators. Cogn Comput 5(2):164–169
7. Deng J, Elberger AJ (2001) The role of pioneer neurons in the development of mouse visual cortex and corpus callosum. Anat Embryol 204(6):437–453
8. Morante-Oria J, Carleton A, Ortino B, Kremer EJ, Fairen A, Lledo P-M (2003) Subpallial origin of a population of projecting pioneer neurons during corticogenesis. PNAS 100(21):12468–12473
9. Shatz CJ, Chun JJM, Luskin MB (1988) The role of the subplate in the development of the telencephalon. In: Jones EG, Peters A (eds) The cerebral cortex. The development of the cerebral cortex, Vol. III. Plenum, New York, pp 35–58
10. Allendoerfer KL, Shatz CJ (1994) The subplate, a transient neocortical structure: Its role in the development of connections between thalamus and cortex. Annu Rev Neurosci 17:185–218
11. Paldino A, Harth E (1977) A computerized study of Golgi-impregnated axons in rat visual cortex. In: Lindsay RD (ed) Computer analysis of neuronal structures. Plenum, New York, pp 189–207
12. Braitenberg V, Schuz A (1998) Cortex: statistics and geometry of neuronal connectivity, 2nd edn. Springer, Berlin
13. Bollobas B, Kozma R, Miklos D (eds) (2009) Handbook of large-scale random networks. Bolyai society mathematical studies. Springer, New York. ISBN: 978-3-540-69394-9
14. Freeman WJ, Kozma R, Bollobas B, Riordan O (2009) Chapter 7. Scale-free cortical planar network. In: Bollobas B, Kozma R, Miklos D (eds) Handbook of large-scale random networks. Series: Bolyai mathematical studies, vol 18. Springer, New York, pp. 277–324
15. Freeman WJ, Quian Quiroga R (2013) Imaging brain function with EEG: advanced temporal and spatial analysis of electroencephalographic and electrocorticographic signals. Springer, New York
16. Zhou S, Mondragon RJ (2004) Accurately modeling the internet topology. Phys Rev E 70(6):066108

17. Zamora-Lopez G, Zhou C, Kurths J (2011) Exploring brain function from anatomical connectivity. Front Neurosci, 5
18. Van den Heuvel MP, Sporns O (2011) Rich-club organization of the human connectome. J Neurosci 31(44):15775–15786
19. Bullmore E, Sporns O (2012) The economy of brain network organization. Nat Rev Neurosci 13(5):336–349

Chapter 6
Modeling Cortical Phase Transitions Using Random Graph Theory

6.1 Describing Brain Networks in Terms of Graph Theory

6.1.1 Synchronization and the 'Aha' Moment

At the microscopic level, the nodes represent neurons and links represent synapses between pairs of neurons. At the mesoscopic level, the nodes represent local populations, and the links represent densities of types of synapses, for example, glutaminergic vs. GABAergic, in local domains such as hypercolumns [1] connected internally and externally by axonal and dendritic bundles. At the macroscopic level the nodes represent modules [2], lobes and Brodmann areas with distinctive cyto-architectures linked by major pathways and tracts [3]. A graph/network at one level is a node at a higher level by integration and condensation and vice versa by differentiation and expansion.

At the microscopic level the nodes are neurons that initially replicate in large numbers, migrate to the surface of the brain, and grow axons and dendrites by which they form synaptic connections. Neurons continue to branch, extend, and form new connections, long after replication and loss of excess neurons by apoptosis (programmed cell death) have ceased. In networks the radial lengths of axons/dendrites from cell bodies determine distances. Most dendritic trees are small enough (<1 mm radius) to sum their currents by passive cable spread; most axons are long enough (>1 mm) to require communication by pulses that use local energy so as to transmit with delay but without attenuation. Every synapse requires a dendritic loop current path to the axon, hence the large surface area of the dendritic tree and the high packing density of neurons ($10^5/\mathrm{mm}^3$) and synapses ($10^9/\mathrm{mm}^3$) to accommodate the 10^4 synapses on each neuron sent by 10^4 other neurons [4–6]. So great is the density of neurons and their connecting fibers that on average a neuron demarcated by its cell body and nucleus connects sparsely with $<1\,\%$ of the neurons within its dendritic arbor, and the likelihood of reciprocal connections between pairs is $<10^{-6}$ [7].

R. Kozma and W.J. Freeman, *Cognitive Phase Transitions in the Cerebral Cortex – Enhancing the Neuron Doctrine by Modeling Neural Fields*,
Studies in Systems, Decision and Control 39, DOI 10.1007/978-3-319-24406-8_6

Neural connections by synapses (not considering gap junctions) are unidirectional; reciprocal interconnections are by different axons. The result is parcellation of the activity of a brain network into a giant component, one or more input components, one or more output components, and miscellaneous islands having insufficient densities of links to participate in the other three groups [8]. The concept of the giant component provides an interesting candidate for a neural correlate of consciousness by virtue of its immense size, unity, and capacity for rapid state transitions [9–11]. The islands, evolving in relative isolation by continuing growth of connections, might be fruitful for using scalp EEG to study the process of insight, the *'Aha'* phenomena described by Jacques Hadamard [12] and as the collision of matrices of thought by Arthur Koestler [13], which has been documented in the auditory cortical EEG of gerbils engaged in category learning [14].

6.1.2 Practical Considerations on Synchrony

Intermittent large-scale synchronization is an important property of intentional brain dynamics as observed by ECoG and EEG experiments. To quantify synchronization effects, we divide the lattice graph into channels using some granulation level. An example of such granulation is shown in Fig. 6.1 by dividing 64 lattice elements into 16 channels. We then compare the activation of each channel with the overall activation of the graph.

When measuring synchrony, first we find the dominant frequency over the entire graph, or ensemble, for the duration of experiment with the discrete Fourier transform.

0	1	2	3	4	5	6	7
8	9	10	11	12	13	14	15
16	10	18	19	20	21	22	23
24	25	26	27	28	29	30	31
32	33	34	35	36	37	38	39
40	41	42	43	44	45	46	47
48	49	50	51	52	53	54	55
56	57	58	59	60	61	62	63

Fig. 6.1 Example of dividing 64 labeled vertices into 16 channels. Edges are not shown

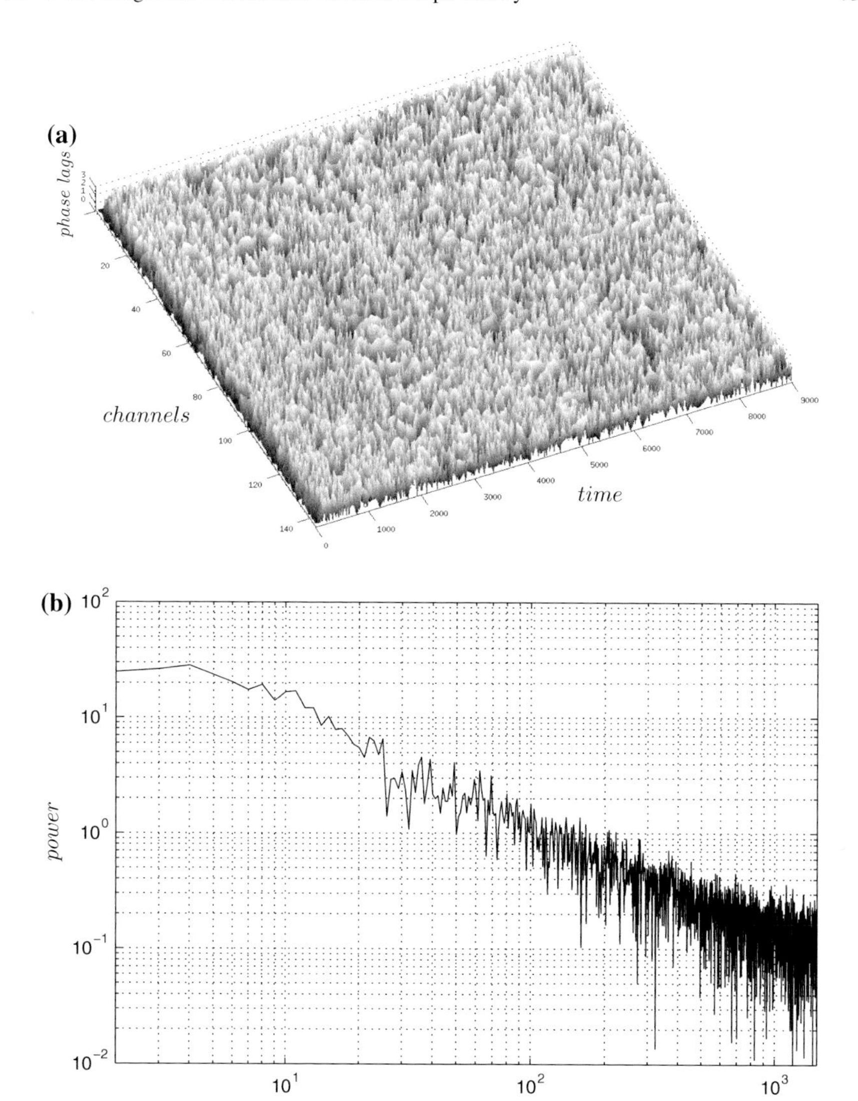

Fig. 6.2 Illustration of phase synchronization results for $G^{2,0}(5)$ with 144 channels. **a** Phase lag values between channels; the channels show random phase relationships, i.e., they are never in synchrony. **b** Power spectral density function in the neighborhood of criticality ($p = 0.14$)

Then, we set a time window W based on the dominant frequency, and the correlations are determined according to this window between the channels and ensemble average. We correlate each channel with the ensemble for duration of W at each time step. Then, the dominant frequency is found for each channel and the ensemble average and their phase shifts are compared. The phase difference at the dominant frequencies measures the phase lag between the channel and the ensemble. The vector composed of phase lags for each channel describes the synchrony of graph components as it evolves in time.

6.1.3 Results of Synchronization Measurements

Results for a single 2D lattice graph are shown in Fig. 6.2a. The phase lag values between channels show random distribution, i.e., the channels do not show synchronous dynamics. Figure 6.2b shows the calculated power spectral densities, which exhibit a scale-free behavior over a wide frequency band of 4 Hz–60 Hz.

6.2 Evolution of Critical Behavior in the Neuropil—a Hypothesis

We propose the following hypothesis on the emergence of critical behavior with the potential of frequent phase transitions in the neuropil between organized an disorganized states [15]. The neural connectivity is sparse in the neuropil at the embryonic stage. The connection density starts increasing even before birth, and approaches a critical level during the first year after birth. Below the critical point, the cortical activity decays to zero after perturbation ($K0$ property). Above the critical connection density, the cortical activity is self-sustaining, indicating the formation of KI population. This creates sustained background noise activity in the cortex. Under continued bombardment by noise, the cortex creates non-random, structured patterns. Near critical states, the brain as a collective system can undergo repeated phase transitions in a self-organized way, under the influence of external and internal factors. We envision the following steps in the ontogenetic development of brains:

- **Start**: Begin with an initial neural population having specific local properties (pioneer neurons).
- **Extensive growth phase**: The long-range connectivity is incrementally increased using some strategy, e.g., preferential attachment, exponential expanding graph (EEGm), or other approaches. The extensive evolution is stopped when criticality is achieved (infant <1 year).
- **Intensive growth phase**: From this stage, the network structure is mostly fixed. Modifications still happen; e.g., due to learning effects through the formation of

Hebbian assemblies, habituation, from infant to child and adolescence, and finally adulthood.

- **Control at the edge**: Using sensory inputs and endogenous noise, the system is balanced near criticality and exhibits phase transitions between highly organized and less-organized phases (adult).
- **Action-perception cycle**: Operating the system through repeated phase transitions as it executes action into the environment, senses the consequences of action, processes, retrieves, and transforms sensory data, compares with anticipated responses, and generates knowledge and meaning.

The above strategy is schematically illustrated in Fig. 6.3. In realistic neural systems, clearly a host of additional factors play a crucial role. However, the mechanism is very robust and it may provide the required dynamical repertoire of behaviors in a wide range of real life conditions. Recent experimental studies indicate that major structural modules of brains were in place by the second year of age; during subsequent development, they continued to strengthen their profile [16]. These results are in line with our hypotheses outlined earlier.

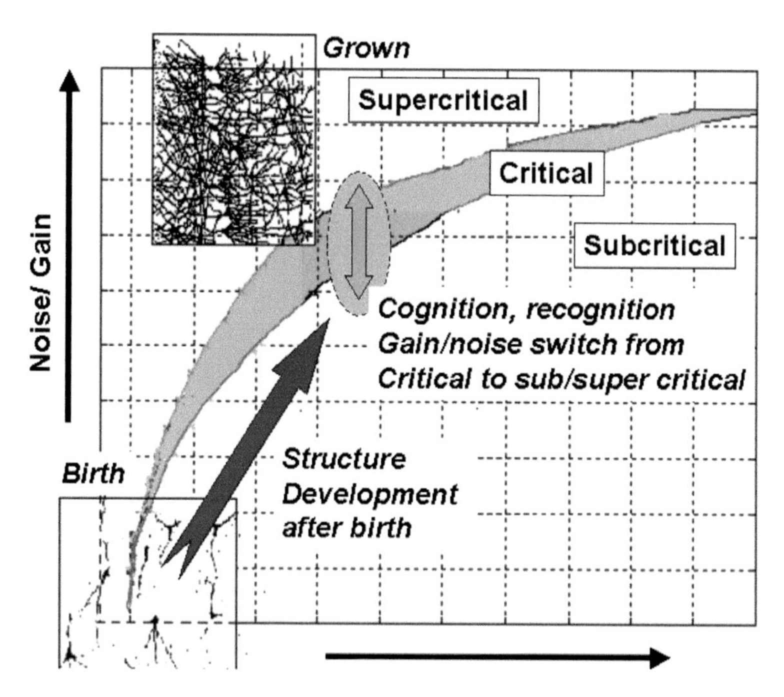

Fig. 6.3 Illustration of self-organization of critical behavior in the percolation model of the neuropil; after [15]

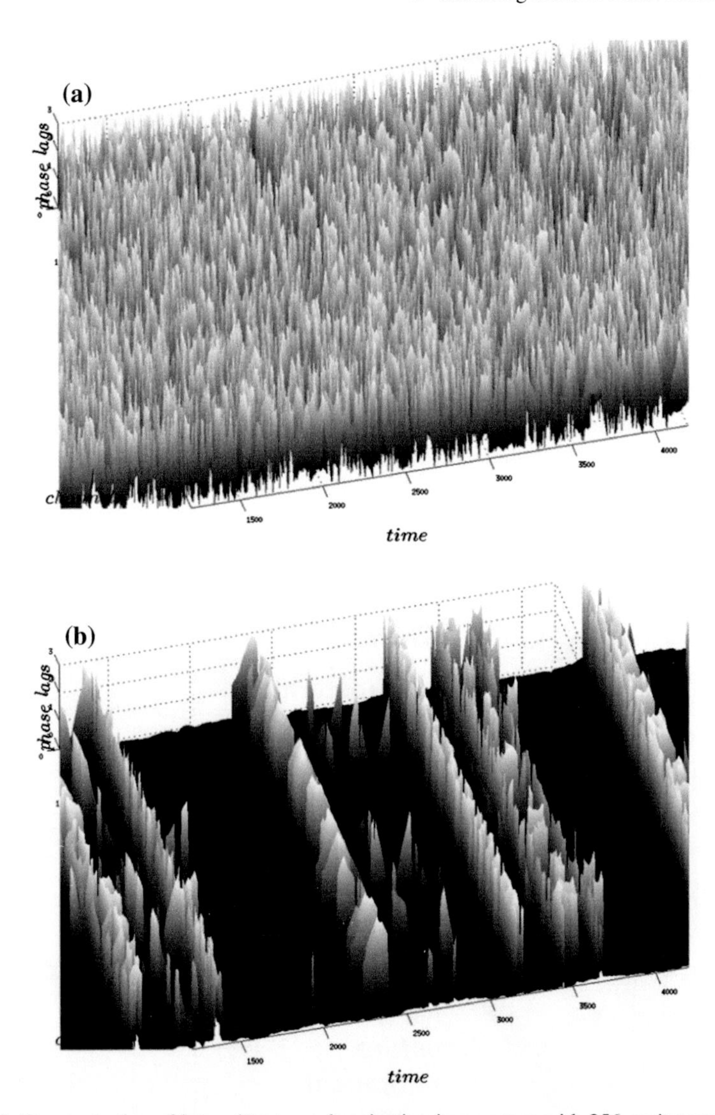

Fig. 6.4 Demonstration of intermittent synchronization in a system with 256 excitatory and 256 inhibitory populations and with mixed local and mean field connections; phase lags are shown for individual channels across time (x axis) and space (y axis). The noise level corresponds to **a** supercritical (p = 0.13) and **b** critical (p = 0.15) levels. There is no synchrony above critical noise, but spontaneous, intermittent synchronization emerges across the array at critical noise level; from [17]

6.3 Singularity and Sudden Transitions—Interpretation of Experimental Findings

We recall the evidences introduced in Part I about singular transitions in cinematic displays of band-pass filtered, high-resolution ECoG images. The dominant large-scale synchrony has been interrupted with brief desynchronization events, indicating the onset of large phase dispersion across the cerebral cortex [18, 19].

Figure 6.4 illustrates the behavior of the KIII model implemented in neuropercolation. The spatial distribution of synchronization at the supercritical regime shows oscillations with high dispersion of the phase, Fig. 6.4a. Near critical parameters, intermittent oscillations emerge, Fig. 6.4b, when relatively quiet periods of high synchrony (black regions) are followed by periods of intensive oscillations with desynchronization (lighter colors) [17, 20]. At subcritical regimes, patterns with high synchronization are observed. The sparseness of connectivity to and from inhibitory populations acts as a control parameter, in addition to the system noise level and the rewiring ratio.

Our neuropercolation model exhibits intermittent synchronization and desynchronization behavior observed in biological brains. Neuropercolation can be considered as a promising novel approach to model large-scale synchronization-desynchronization of brain functions. Future work aims at studying the effect of refractoriness, which is expected to help to stabilize the activity of excitatory populations, the role of time delays between cortical regions, as well as analyzing the spatial structure of amplitude and phase modulations.

References

1. Mountcastle VB (ed) (1974) Medical Physiology, 13th edn. Saint Louis, C.V. Mosby Co., Missouri, p 232
2. Houk JC (2005) Agents of the mind. Biol Cybern 92(6):427–437
3. Brodmann K (1909) Vergleichende Lokalizationslehre der Grosshirnrinde. Barth, Leipzig
4. Sholl DA (1956) The organization of the cerebral cortex. Wiley, London
5. Bok ST (1959) Histonomy of the cerebral cortex. Elsevier, Amsterdam
6. Braitenberg V, Schuz A (1998) Cortex: statistics and geometry of neuronal connectivity, 2nd edn. Springer, Berlin
7. Thompson H, (1899) The total number of functional nerve cells in the cerebral cortex of man. J Comp Neurol 9:113–140 (given as 9,282,826,403 which Warren McCulloch (1967) rounded off to 10^{10})
8. Nelson SM, Cohen AL, Power JD et al (2010) A parcellation scheme for human left lateral parietal cortex. Neuron 67(1):156–170
9. Werner G (2007) Metastability, criticality, and phase transitions in brains and its models. BioSystems 90(496–508):2007
10. Werner G (2007) Brain dynamics across levels of organization. J Physiol Paris 101:273–279
11. Wallace R (2007) Culture and inattentional blindness: a global workspace perspective. J Theor Biol 245(2):378–390
12. Hadamard J, (1945) The mathematician's mind: the psychology of invention in the mathematical field. Princeton University Press, New Jersey

13. Koestler A (1964) The act of creation. Macmillan, New York
14. Ohl FW, Scheich H, Freeman WJ (2001) Change in pattern of ongoing cortical activity with auditory category learning. Nature 412:733–736
15. Kozma R, Puljic M, Balister P, Bollobas B, Freeman WJ (2005) Phase transitions in the neuropercolation model of neural populations with mixed local and non-local interactions. Biol Cybern 92:367–379
16. Hagmann P, Cammoun L, Gigandet et al (2008) Mapping the structural core of human cerebral cortex. PLoS Biol 6(7) e159 :1–14
17. Kozma R, Puljic M (2013) Hierarchical random cellular neural networks for system-level brain-like signal processing. Neural Netw. 45:101–110
18. Kozma R, Freeman WJ (2008) Intermittent spatio-temporal desynchronization and sequenced synchrony in ECoG signals. Chaos 18:037131
19. Freeman WJ, Vitiello G (2010) Vortices in brain waves. Int J Mod Phys B 24(17):3269–3295
20. Puljic M, Kozma R (2010) Broad-band oscillations by probabilistic cellular automata. J Cell Autom 5(6):491–507

Chapter 7
Summary of Main Arguments

7.1 Brain Imaging Combining Structural and Functional MRI, EEG, MEG and Unit Recordings

The hypothesis that foci that are visualized with fMRI are signs of hubs rather than modules can be tested by combining hemodynamic imaging [1] with EEG imaging [2, 3] and MEG [4]. Experimental data indicate that the necessary macroscopic frames with beta-gamma carrier frequencies are readily found in human volunteers engaged in cognitive tasks by several research groups [5–11]. Strong relations of beta and gamma oscillations to theta have been identified in scalp EEG [12–18].

Single neurons in human and animal neocortex generate spike trains while performing precise cognitive tasks, for example, face recognition [19]. These cells act as higher-order feature detector neurons [20], synfire chains [21, 22], colloquially called *grandmother cells* [23]. If these neurons were randomly distributed in the neocortex, the likelihood of finding any of them in brains would be vanishingly small. The fact that they can be found at all implies that in brains they are concentrated, possibly in distributed Hebbian cell assemblies that enact a designated task [24]. The existence of such distributed assemblies is in line with a holistic view of brain functions [25].

7.2 Significance of RGT for Brain Modeling

Random graph and percolation theory teaches us that, in a regular lattice structure, exponentially long time is required to achieve conditions for phase transitions, in terms of the lattice size. On the other hand, once the conditions are ready for a switch in the spatial pattern, the transition happens rapidly, at polynomial time scale. Hence the terminology, neuropercolation is justified. We observe frame repetition rates at the theta-beta band with average time of 100–250 ms, while the transition

R. Kozma and W.J. Freeman, *Cognitive Phase Transitions in the Cerebral Cortex – Enhancing the Neuron Doctrine by Modeling Neural Fields*,
Studies in Systems, Decision and Control 39, DOI 10.1007/978-3-319-24406-8_7

happening at a much shorter time scale (in about 10 ms). Our theoretical results show that brains have a very efficient mechanism to generate the conditions for very rapid switches and reduce the waiting time between the transitions, by using long axons for long-range effects across mesoscopic cortical distances [26, 27].

7.2.1 Relevance to Brain Diseases

Network theory indicates that preferential attachment evolution rule can lead to the development of hub structure. Such networks are relatively unaffected by random loss of nodes but catastrophically fail under targeted attack aiming at hubs [28]. It is a highly disputed issue, whether brains have scale-free structures with hubs, most of the existing evidence indicates otherwise. Functionally, however, the existence of hubs has been demonstrated under various cognitive conditions [5, 29]. The performance of cerebral cortex is relatively resistant to degradation by local lesions in areas not affecting input and output pathways. The medical literature is rich with descriptions of transient loss of circumscribed cognitive functions that accompany focal damage to limited areas of neocortex [30] and are repaired by learning, but cortex can fail catastrophically with certain critical lesions, such as damage to the midbrain reticular formation leading to coma, the substantia nigra leading to Parkinson's disease, and the perforant path in Alzheimer's disease. Functional evidence for hubs might be seen in the patches of cortical activity seen in fMRI that are associated with a broad range of cognitive functions [1].

Significant preferentially is seen in the mechanism by which pioneer neurons provide the initial guidance that is required for the construction of topographic maps, which are displayed in textbooks as motor and sensory homunculi. This mechanism addresses the local organization that is required for sensorimotor integration by neurons in Layers II to IV. The question is whether another form of preferentiality might be modeled in Layers I, V and VI, which might help to explain hubs, if indeed they exist. Our theoretical model based on the pioneer neurons role in development indicated scale-free length distribution and exponential degree distribution [31].

Considering the contradictory experimental results on scale-free brain network properties, we hypothesize that brains exhibit structural properties resembling scale-free distributions, thus benefiting from the robustness provided against random damage, but also include certain additional properties deviating from the strict power-law. Namely, the hub structure may include certain level of overlaps to avoid the shortcomings of having catastrophic consequences due to loss of hubs. The requirements for macroscopic connectivity supporting the resistance of the network to loss of hubs in neocortex are unknown. A mathematical model for large-scale evolutionary rules is needed to guide anatomists and physiologists in the search for the requisite structural and functional properties of cortex. Such results may give a fresh start to understanding the enigmas of speech localization, and the paradox that the higher is the verbal IQ, the smaller is the speech area [32]. Broca's area is in close conjunction with the motor areas for the lips and tongue, yet linked by the uncinate fasciculus to the temporal lobe for audition and by U-fibers to the parietal lobe for proprioception. Wernicke's area is at the conjunction of somatic, visual and auditory sensory cortices. In both instances the tendencies for preferential connectivity may support to some degree the growth of hubs respectively for motor and perceptual speech functions.

Focal electrical or magnetic pulse stimulation by neurosurgeons temporarily inactivates speech, but that does not imply that speech is localized to the site of stimulation. Instead the elucidation of speech mechanisms requires investigation of how neurons in a hub participate in macroscopic coordination of oscillatory neural activity. Definitive exploration of hubs and their partial overlap may be optimized by recording EEG/MEG from a dense array of channels overlying a large percentage of the surface each hemisphere, and with sampling at 1 cm intervals corresponding to the width of gyri so as to capture adequately the texture of AM patterns of spatially coherent activity [33]. A foreseeable device with 64 channels for EEG in the form of a linear array of 30 cm long would comfortably sit over the forehead.

7.2.2 Neuropercolation as a Novel Mathematical Tool

Neuropercolation replaces differential equations with probabilistic cellular automata motivated by the properties of the cerebral cortex. Phase transitions model transient effects and metastable states observed in brain dynamics [34–37] and they are deemed crucial components of the language of the brain as envisioned by Von Neumann [38]. The main conclusions regarding the modeling of experimentally observed collective neural field effects and hypotheses for future studies are summarized below:

- PCA and neuropercolation theory teach us that, in large-scale networks, metastable activity patterns can be maintained for some time, but these patterns will inevitably collapse through a phase transition at a rate of several patterns per second, leading to the emergence of new patterns. Long axons communicating across mesoscopic cortical distances can effectively control the extent of the waiting time between the transitions and the rapid switching from one metastable AM pattern to another. Phase transitions in the neuropil can be interpreted in terms of deviations from scale-free distributions during the emergence of large-scale catastrophic events [27, 39], such as *Dragon Kings* [40, 41].
- Neuropercolation helps to identify specific network structures in brains that support robust operation. In our exponentially expanding graph model (EEGm), pioneer neurons facilitate the formation of a highly connected core [31, 42]. We hypothesize that the highly connected core produced by pioneer neurons in neuropercolation and the *Rich Club* property observed in experimental data are key preconditions of controllable phase transitions. RCs help to prevent the catastrophic loss of functions if a specific hub is damaged. Brains may implement the ingenious design of multiple hubs with overlapping functions such as RCs, to minimize consequences of any damage or anomaly.

- We propose the hypothesis that the rapidly propagating phase gradients observed across the cerebral cortex in the form of phase cones [43] originating at a given point of space and time can been linked with the *'Aha'* moment and manifested in singularity of the neurodynamics in the cerebral cortex [44, 45]. Extensive studies are underway to validate this hypothesis.

The challenge of describing and modeling the complexities of immense numbers of neurons and synapses in the cortex has been met by invoking a continuum to represent the mesoscopic state variables of neuronal populations in networks of ordinary differential equations [46]. This strategy works well with piece-wise linearization giving analytic solutions to the equations. Networks of populations with several types of nonlinear feedback cannot be solved analytically but require numerical integration. Owing to attractor crowding and numerical instabilities from the extremely large number of attractors in high-dimensional networks, the digital models are unstable and must be tempered with additive noise [47, 48] leading to stochastic difference equations. These equations suffice to replicate the time series and spatial patterns of wave packets governed by nonconvergent attractors before and after phase transitions [49], however, they are unsuitable to describe sudden transitions from one state to another.

Random graph theory offers a fresh beginning, when the discreteness of neurons and synapses can be approximated in percolation theory, which has been used to describe forest fires and epidemics. It is readily adapted to describe neural nets at the microscopic level, and to describe nodes at mesoscopic and macroscopic levels. Neuropercolation theory is already well advanced in modeling structural and functional connectivity in the cortex [27, 50] especially in modeling the role of long connections, inhibitory feedback, and noise in the genesis of self-regulated spontaneous activity of large nets of nodes at the mesoscopic level. Results to date suffice to simulate white noise with a flat PSD at low levels of interaction strength, brown noise with $1/f^2$ PSD when close to a phase boundary, and narrow band gamma oscillation just beyond a state transition. The tools of random graph theory, applied in the context of the developmental neuroanatomy of cortex to the electrophysiological signs of activity recorded in trained subjects, can resolve conflicting evidence for local vs. holistic information processing in brains by explaining how large-scale fields of activity emerge and coordinate local, modular activities.

Definitive exploration of brain networks may be optimized by recording EEG/MEG from a wide array of channels overlying each hemisphere, and with sampling at 0.5 cm intervals determined by the width of gyri so as to capture adequately the texture of AM patterns of spatially coherent activity [51]. Devices on this scale are under development; a foreseeable device with 48 channels for EEG in the form of a 20 cm long linear array would comfortably sit over the forehead [52].

7.3 Neuromorphic Nanoscale Hardware Platforms

To achieve parallelism for the task ahead, we partition the set of vertices into multiple components. The most efficient way is to partition vertices into a number of subsets that is equal the number of available computer processors. Vertices and their neighborhoods are stored in shared memory, which is divided so that each piece holds one component. A piece of memory is accessed by one processor, which evolves the set of vertices. For each vertex at each time step, the vertex neighborhood is checked to determine the vertex's state. When the processor finishes checking the (majority) rule of all vertices, it is paused to wait until all the processors finish their evaluation. When all the processors have finished their task on their vertices, a new cycle begins. This synchronizes the evolution of the entire graph.

Cellular automata with some level of rewiring have enormous potential for modeling complex systems. A prominent example of such architecture is the cellular neural network (CNN) [53]. Cellular networks can be viewed as models of a spatially decentralized system made up of a number of individual processors. The direct communication between processors is limited to the interactions in the processor neighborhood. Such machines have a high degree of parallelism. Fastest implementation of graph evolution is possible with new hardware in which there would be many simpler processors and each would implement the majority rule and locally store the state of vertices. Each processor would also be directly connected or wired to its neighborhood of processors.

Subset of processors, or the hardware vertices, would be allocated for the output of calculations of system as a whole. Because of their simplicity, regularity, and the option of hard-wired neighborhoods which are not too large, cellular automata are natural candidates for VLSI implementation. In locally connected cellular automata, circuit design reduces to the design of a single simple cell, and layout is uniform; the whole mask for a large cellular-automaton array can be generated by a step-and-repeat procedure; essentially no silicon area is wasted on long interconnection lines. Our neuropercolation network models neuronal populations and is suitable for VLSI implementation; see Fig. 7.1.

Neuromorphic considerations indicate that some level of non-local connectivity is required for robust operation in large-scale networks. The rewiring ratio in cortical neurons is below 0.1 %. This level of non-locality is considered affordable by advanced Field Programmable Gate Array (FPGA) architectures [55]. The mixture of mostly short connections with a small amount of long links gives rise to a new generation of parallel and distributed machines [54]. Alternative approaches pursue analog platforms to benefit from the intrinsic noise in such systems, combined with memristive properties [56]. A potential neuromorphic embodiment includes Atomic Switch Networks (ASN) [57, 58] in very dense *nanoscale networks* consisting memristive compounds. Intensive studies are underway to develop a comprehensive approach to neuromorphic hardware for neuropercolation models based on nano-scale hardware platforms.

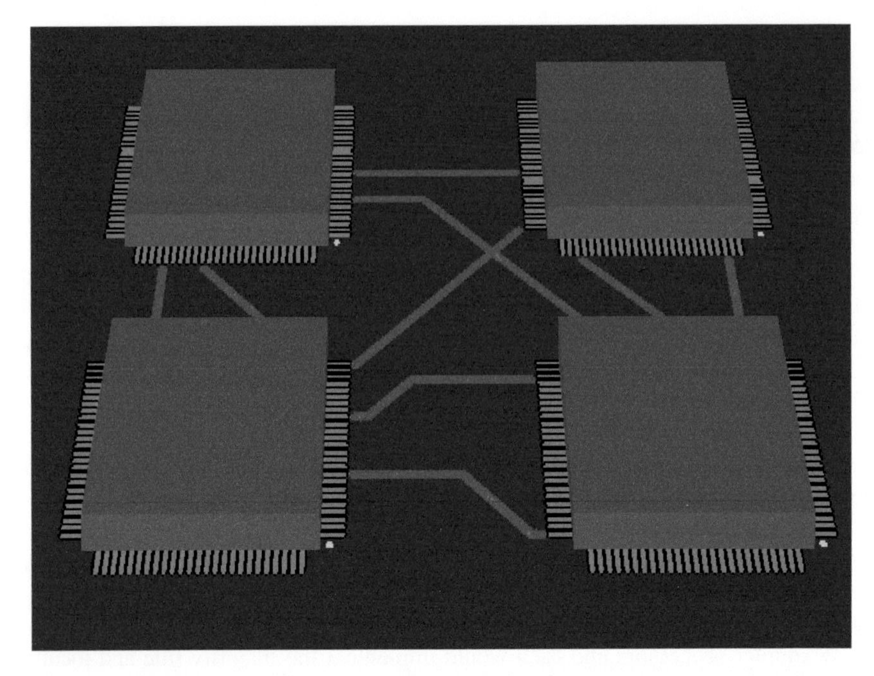

Fig. 7.1 Illustration of a potential hardware approach for the implementation of a Freeman KIII network using neuropercolation model using four coupled cellular network chip arrays with sparse interconnections [54]

References

1. Buxton RB (2001) Introduction to functional magnetic resonance imaging: principles and techniques. Cambridge University Press, Cambridge
2. Barlow JS (1993) The electroencephalogram: its patterns and origins. MIT Press, Cambridge
3. Pfurtscheller G, Lopes da Silva FH (eds) (1988) Functional brain imaging. Hans Huber Publishers, Lewiston
4. Hamalainen M, Hari R, Ilmoniemi RJ, Knuutila J, Lounasmaa OV (1993) Magnetoencephalography—theory, instrumentation, and applications to noninvasive studies of the working human brain. Rev Mod Phys 65:413–497
5. Reijneveld JC, Ponten SC, Berendse HW, Stam CJ (2007) The application of graph theoretical analysis to complex networks in the brain. Clin Neurophysiol 118(11):2317–2331
6. Stam CJ, Jones BF, Nolte G, Breakspear M, Scheltens P (2007) Small-world networks and functional connectivity in Alzheimer's disease. Cereb Cortex 17:92–99
7. Stam CJ (2010) Characterization of anatomical and functional connectivity in the brain: a complex networks perspective. Int J Psychophysiol 77(3):186–194
8. Bressler S, Menon V (2010) Large-scale brain networks in cognition: emerging methods and principles. Trends Cogn Sci 14:277–290
9. Bullmore E, Sporns O (2012) The economy of brain network organization. Nat Rev Neurosci 13(5):336–349
10. Kozma R, Freeman WJ (2014) On neural substrates of cognition: theory, experiments and application in brain computer interfaces. In: IEEE biomedical science and engineering center conference (BSEC), Annual Oak Ridge National Laboratory, pp 1–4

11. Hillebrand A, Stam CJ (2014) In magnetoencephalography. Recent developments in MEG network analysis. Springer, Berlin, pp 263–277
12. Freeman WJ, Burke BC, Holmes MD (2003) Aperiodic phase re-setting in scalp EEG of beta-gamma oscillations by state transitions at alpha-theta rates. Hum Brain Mapp 19(4):248–272
13. Demiralp T, Bayraktaroglu Z, Lenz D, Junge S, Busch NA, Maess B, Herrmann CS (2007) Gamma amplitudes are coupled to theta phase in human EEG during visual perception. Int J Psychophysiol 64(1):24–30
14. Ramon C, Freeman WJ, Holmes M, Ishimaru A, Haueisen J, Schimpf PH, Rezvanian E (2009) Similarities between simulated spatial spectra of scalp EEG MEG and structural MRI. Brain Topogr 22(3):191–196
15. Berthouze L, James LM, Farmer SF (2010) Human EEG shows long-range temporal correlations of oscillation amplitude in theta, alpha and beta bands across a wide age range. Clin Neurophysiol 121(8):1187–1197
16. Van de Ville D, Britz J, Michel CM (2010) EEG microstate sequences in healthy humans at rest reveal scale-free dynamics. Proc Natl Acad Sci 107(42):18179–18184
17. Ruiz Y, Pockett S, Freeman WJ, Gonzales E, Guang Li (2010) A method to study global spatial patterns related to sensory perception in scalp EEG. J Neurosci Methods 191:110–118
18. Freeman WJ (2015) Perspectives: mechanism and significance of global coherence in scalp EEG. In: Buzsaki G, Freeman WJ (eds) current opinion in neurobiology 31: brain rhythms and coordination, pp 199–207
19. Quian Quiroga R, Reddy L, Kreiman G, Koch C, Fried I (2005) Invariant visual representation by single-neurons in the human brain. Nature 435:1102–1107
20. Singer W, Gray CM (1995) Visual feature integration and the temporal correlation hypothesis. Annu Rev Neurosci 18:555–586
21. Abeles M (1991) Corticonics: neural circuits of the cerebral cortex. Cambridge University Press, New York
22. Mehring C, Hehl U, Kubo M, Diesmann M, Aertsen A (2003) Activity dynamics and propagation of synchronous spiking in locally connected random networks. Biol Cybern 88:395–408
23. Lettvin JY (1995) J Y Lettvin on grandmother cells. In: Gazzaniga MS (ed) The cognitive neurosciences. MIT Press, Cambridge
24. Freeman WJ, Quian Quiroga R (2013) Imaging brain function with EEG: advanced temporal and spatial analysis of electroencephalographic and electrocorticographic signals. Springer, New York
25. Pribram KH (1991) Brain and perception: holonomy and structure in figural processing. Psychology Press, New York
26. Balister P, Bollobas B, Johnson JR, Walters M (2010) Random majority percolation. Random Struct Algorithms 36(3):315–340
27. Kozma R, Puljic M (2015) Random graph theory and neuropercolation for modeling brain oscillations at criticality. Curr Opin Neurobiol 31:181–188
28. Wang XF, Chen G (2002) Synchronization in scale-free dynamical networks: robustness and fragility. IEEE Trans Circuits Syst Fund Theory Appl 49:54–62
29. Bassett DS, Meyer-Lindenberg A, Achard S, Duke T, Bullmore E (2006) Adaptive reconfiguration of fractal small-world human brain functional networks. PNAS 103(51):19518–19523
30. Critchley EM (1979) Drug-induced neurological disease. BMJ 1(6167):862–865
31. Freeman WJ, Kozma R, Bollobas B, Riordan O (2009) Scale-free cortical planar network. In: Bollobas B, Kozma R, Miklos D (eds) In: Handbook of large-scale random networks, Bolyai Mathematical Studies. Springer, New York, pp 277–324
32. Ojemann GA (2003) The neurobiology of language and verbal memory: observations from awake neurosurgery. Int J Psychophysiol 48(2):141–146
33. Freeman WJ (2003) The wave packet: an action potential for the 21st century. J Integr Neurosci 2:3–30
34. Werner G (2007) Metastability, criticality, and phase transitions in brains and its models. BioSystems 90(496–508):2007

35. Steyn-Ross DA, Steyn-Ross ML (eds) (2010) Modeling phase transitions in the brain. Springer series computational neuroscience, vol 4. Springer, Berlin
36. Rabinovich MI, Friston KJ, Varona P (eds) (2012) Principles of brain dynamics. MIT Press, Cambridge
37. Tognoli E, Kelso JAS (2014) The metastable brain. Neuron 81(1):35–48
38. Von Neumann J (1958) The computer and the brain. Yale University Press, New Haven
39. Erdi P, Kozma R, Puljic M, Szente J (2013) Neuropercolation and related models of criticalities. In: European meeting of statisticians Hungary contents, p 106
40. Sornette D, Quillon G (2012) Dragon-kings: mechanisms, statistical methods and empirical evidence. Eur Phys J Spec Top 205(1):1–26
41. Pisarenko VF, Sornette D (2012) Robust statistical tests of Dragon-Kings beyond power law distributions. Eur Phys J Spec Top 205(1):95–115
42. Freeman WJ (2009) Deep analysis of perception through dynamic structures that emerge in cortical activity from self-regulated noise. Cogn Neurodyn 3(1):105–116
43. Freeman WJ, Holmes MD, West GA, Vanhatalo S (2006) Fine spatiotemporal structure of phase in human intracranial EEG. Clin Neurophysiol 117:1228–1243
44. Freeman WJ (2007) Proposed cortical "shutter" mechanism in cinematographic perception. In: Perlovsky L, Kozma R (eds) Neurodynamics of cognition and consciousness. Springer, Heidelberg, pp 11–38
45. Kozma R, Davis JJ, Freeman WJ (2012) Synchronization of de-synchronization events demonstrate large-scale cortical singularities as hallmarks of higher cognitive activity. J Neurosci Neuro-Eng 1(1):13–23
46. Freeman WJ (1975/2004) Mass action in the nervous system. Academic Press, New York. Electronic version 2004—http://sulcus.berkeley.edu/MANSWWW/MANSWWW.html
47. Freeman WJ, Chang H-J, Burke BC, Rose PA, Badler J (1997) Taming chaos: stabilization of aperiodic attractors by noise. IEEE Trans Circuits Syst 44:989–996
48. Kozma R (2003) On the constructive role of noise in stabilizing itinerant trajectories on chaotic dynamical systems. Chaos 11(3):1078–1090
49. Kozma R, Freeman WJ (2001) Chaotic resonance: methods and applications for robust classification of noisy and variable patterns. Int J Bifurc Chaos 10:2307–2322
50. Kozma R, Puljic M, Balister P, Bollobas B, Freeman WJ (2005) Phase transitions in the neuropercolation model of neural populations with mixed local and non-local interactions. Biol Cybern 92:367–379
51. Freeman WJ, Burke BC (2003) A neurobiological theory of meaning in perception. Part 4. multicortical patterns of amplitude modulation in gamma EEG. Int J Bifurc Chaos 13:2857–2866
52. Kozma R, Freeman WJ, Lin CT (2013) Optimizing EEG/EMG signal/noise ratio, Society for Neuroscience, Abstract 6555, San Diego, 9–13 Nov, San Diego
53. Chua LO, Roska T (1993) The CNN paradigm. IEEE Trans Circuits Syst I: Fundam Theory Appl 40(3):147–156
54. Kozma R, Puljic M (2013) Hierarchical random cellular neural networks for system-level brain-like signal processing. Neural Netw 45:101–110
55. Srinivasa N, Cruz-Albrecht JM (2012) Neuromorphic adaptive plastic scalable electronics: analog learning systems. Pulse, IEEE 3(1):51–56
56. Kozma R, Pino R, Pazienza G (eds) (2012) Advances in neuromorphic memristor science and applications. Springer, Heidelberg
57. Sillin HO, Aguilera R, Shieh HH, Avizienis AV, Aono M, Stieg AZ, Gimzewski JK (2013) A theoretical and experimental study of neuromorphic atomic switch networks for reservoir computing. Nanotechnology 24(38):384004
58. Stieg AZ, Avizienis AV, Sillin HO, Aguilera R, Shieh HH, Martin-Olmos C, Gimzewski JK (2014) Self-organization and emergence of dynamical structures in neuromorphic atomic switch networks. Memristor networks. Springer International Publishing, Switzerland, pp 173–209

Part II
Supplementary Materials on Brain Structure and Dynamics

Chapter 8
Supplement I: Mathematical Framework

8.1 ODE Implementation of Freeman K Sets

8.1.1 Foundations of Freeman K Sets

We developed a dynamical systems approach to spatio-temporal neurodynamics. The corresponding mathematical objects are called Freeman K sets, which are mesoscopic models representing an intermediate-level between microscopic neurons and macroscopic brain structures. K sets are multi-scale models, describing increasing complexity of structure and dynamical behaviors. The dynamics of K set are modeled using a system of nonlinear ordinary differential equations with distributed parameters. K sets describe spatial-temporal patterns of phase and amplitude oscillations, generated by neural populations at each level. They model observable fields of neural activity comprising EEG, LFP, and MEG.

Functionally, K sets describe cortical sensory processes leading to transmission of macroscopic output. The hierarchy of pathways is schematically illustrated in Fig. 8.1. As in all sensory systems, the input pathway to the olfactory bulb (the primary olfactory nerve) is topographically organized, but the output pathway (the lateral olfactory tract) is not. Each transmitting mitral cell diverges its output to a wide distribution of pyramidal neurons in the olfactory cortex, and each cortical neuron receives input from a broad distribution of mitral cells. This divergent-convergent pathway performs a spatial integral transformation on bulbar output. This operation enhances the signal-to-noise ratio of the endogenous bulbar output and attenuates exogenously evoked output. It simultaneously sends the same signal to all parts of the olfactory system and many parts of the limbic system, so it effectively interfaces the receptor array determined by its intrinsic principles of organization with the multiple central neural systems that depend on the abstracted and categorized olfactory information provided by the receptors.

R. Kozma and W.J. Freeman, *Cognitive Phase Transitions in the Cerebral Cortex – Enhancing the Neuron Doctrine by Modeling Neural Fields*,
Studies in Systems, Decision and Control 39, DOI 10.1007/978-3-319-24406-8_8

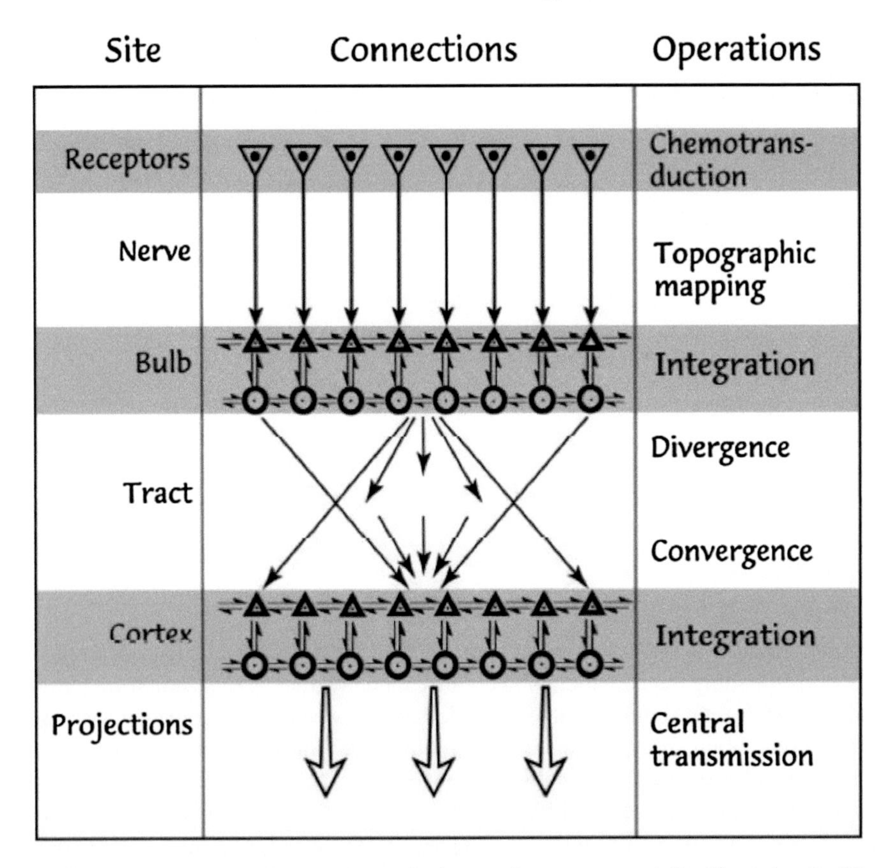

Fig. 8.1 The organization of axon pathways for input and output are specialized for their tasks. The input path delivers microscopic pulses carrying precisely localized spatial and temporal information. The lateral interconnections among mitral cells provide categorization of mesoscopic patterns carried by randomly varying input by Hebbian learning. The inhibitory interneurons by negative feedback to excitatory neurons generate carrier waves that sustain macroscopic patterns. The output path cleans and distributes the same output to multiple targets, with resolution in proportion to the fraction of the total output that each target receives; from [1], Fig. 15, p. 84

K sets were introduced in the 70s by Freeman, named in the honor of Aharon Katzir-Katchalsky, an early pioneer of neural dynamics [2, 3]. The approach represented by K sets is motivated by the idea of modeling populations of neurons. Anatomically, the existence of neural populations is manifested in the micro-column structure of cortical tissues. The basic unit of neural populations corresponding to this granulation level is called K0 set. The number of neurons in the mammalian brain can range from about 10^9 in the case of small animals and approx. 10^{11} in humans. The granulation level of K0 sets can include approx. 10^4 neurons, i.e., it represents an intermediate level between microscopic neurons and the whole brain. The intermediate level is often referred to as mesoscopic level, and the corresponding processes manifest mesoscopic brain dynamics [4]. At the next level of hierarchy we have KI sets. The interaction of mutually excitatory neural populations is modeled by excitatory *KI* set (KI_e). Similarly, the interaction of inhibitory neural populations is modeled by inhibitory *KI* set (KI_i). Coupled systems including the interaction of both excitatory and inhibitory populations are called *KII*, which may exhibit narrow-band oscillatory behavior due to negative feedback. *KII* sets are basic oscillatory units of the K hierarchy. The interaction of several *KII* units with different dominant frequencies produce the *KIII* set.

KIII sets are complex dynamical systems modeling classification in various cortical areas, having typically hundreds of degrees of freedom. In early applications, *KIII* sets exhibited extreme sensitivity to model parameters, which prevented their broad use in practice [5]. Further systematic analysis identified regions of robust performance of *KIII* ([6], and stability properties of K sets have been studied [7, 8]). The utility of K sets has been demonstrated in a wide range of applications, including detection of chemicals [9], classification [10–12], time series prediction [13, 14], navigation [15, 16]. Recent developments include the KIV sets [17–19] for modeling intentional behavior. They are applicable to goal-oriented navigation and robot control [20, 21]. Finally, at the top level is the KV, which models the neocortex as a single organ as postulated by Williams James over a century ago [22].

8.1.2 Hierarchy of Freeman K Sets

During the last decades, extensive research has been conducted in the field of neuroscience which resulted in a thorough understanding of neural processes [23, 24]. Individual neurons are typically modeled using Hodgkin–Huxley equations. The Hodgkin–Huxley and related equations employ a system of first order ordinary differential equations (ODEs) to describe the physical properties of the cell membrane and the concentrations of different ions in the proximity of the membrane. In these models, the state variables are the membrane potentials of the individual neurons and the action potentials (also called pulses or spikes) are determined by solving first-order differential equations with respect to the membrane potentials; see for example [25].

The first order approximation used in these ODEs produces a very efficient way of describing the dynamics of neural pulses using exponential decay in time. In some cases, however, a higher-order approximation may be beneficial, which would allow a dynamical characterization which is more refined than a sequence of pure exponential decays. In the field of physics, the use of second order equations is a common practice, starting from Newton's basic equation of motion. It has been Freeman's groundbreaking insight over 50 years ago, that the description of the dynamics of large masses of neurons requires field approach which can benefit from higher order ODE models [26]. His original idea has lead to the hierarchy of K models studied in this paper. In the next section we introduce the $K0$ set, which is the starting point of the neuron population model.

*K*0 Sets—Pulse to Wave Conversion Dynamics in Basic Neural Populations

The state variables describing neural populations are the averaged wave and pulse densities. They are directly related to the average local field potentials, which can be measured by intracranial EEG electrodes. The wave density relates to the membrane potentials of the dendrites, and the pulse density relates to the action potentials on the axons. As the neurons constantly engage in wave to pulse transformation, it is sufficient to use either of them as state variable, while the other is inferred using the corresponding transformation from wave to pulse or vice versa.

We use the average pulse density of a neural population as the state variable. Let $y_i(t)$ denote the normalized pulse density of the ith neural population $i = 1, \ldots, N$, where N is the number of populations. The normalization is done by making the steady state pulse density equal to zero. Accordingly, y_i can assume positive and negative values as time evolves. The mathematical model of neural population has been originally developed based on intracranial EEG measurements in animals. The results introduced below have been obtained in cats with implanted electrodes in the prepyriform cortex and having injected with a surgically anesthetizing dose of pentobarbital [26]. In this way the parameters of the open-loop transfer function of the prepyriform cortex have been estimated. The results give a second order system with the following components: a simple exponential decay term, and a term describing the delay effect due to distributed lag in the cortical tissue. The experiments indicated the presence of a third exponential term as well, which is neglected for the sake of simplicity. This leads to the following second order ODE for the normalized pulse density $p(t)$ of a neural population, which is called $K0$ equation [26]:

$$(ab)\frac{d^2p(t)}{dt^2} + (a+b)\frac{dp(t)}{dt} + p(t) = F(t). \tag{8.1}$$

Here a and b are biologically determined time constants. p(t) denotes the activation of the node as function of time. $F(t)$ includes a nonlinear sigmoid function $Q(x)$, which acts on the weighted sum of activations from neighboring nodes and external inputs. Freeman points out the close parallel between the population response described by the K0 equation and the postsynaptic potentials generated by single neurons in response to impulses. The population time constants must be consistent with the cumulative effects of synaptic delays, dispersion, passive dendritic conduction, passive decay of membrane transients, dispersal of transmitter substances, etc., but they cannot be identified with unique processes at the cellular level [27].

Unlike the sigmoid curves commonly used in neural network research, $Q(v)$ is asymmetric modeled from experiment on biological neural activation and it is given by the equation [26]:

$$p = Q(v) = q_m \left\{ 1 - exp(-\frac{1}{q_m(e^v - 1)}) \right\} \tag{8.2}$$

In these equations, q_m is the parameter specifying the slope and maximal asymptote of the curve. The value of constant q_m varies between 1 and 14 for different

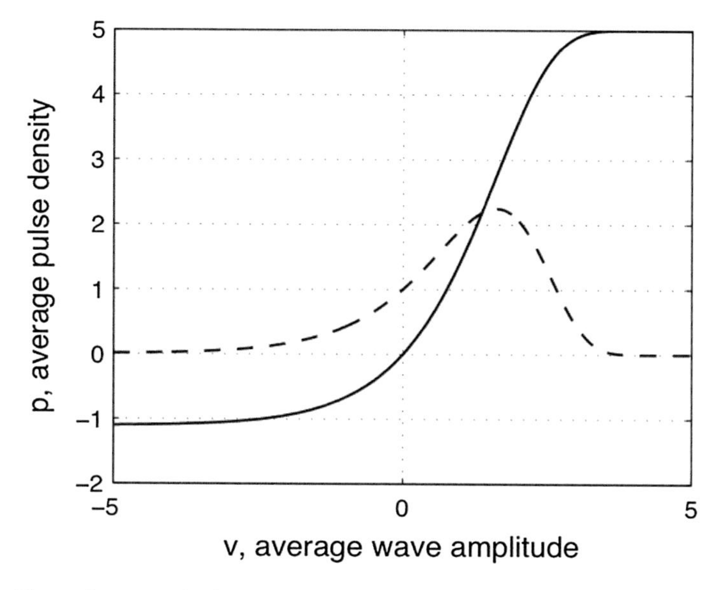

Fig. 8.2 The nonlinear transfer function $Q(v)$ given by Eq. 8.2 (*solid line*) and its derivative (*dashed line*). The maximum of the derivative is shifted to the positive value of the average wave amplitude; $q_m = 5$

types of neural populations and for different states of the animal in sleep or being awake and motivated. The transfer function is depicted in Fig. 8.2 for a value $q_m = 5$, which is a typical value for waking animals [3]. Unlike the sigmoid curves commonly used in neural network research, $Q(v)$ is asymmetric with the level of asymmetry dictated by the constant q_m and its maximum gain is shifted to the positive values of wave amplitudes. The sigmoid function shows that population outputs are bounded by a threshold during inhibition, and by refractory periods during excitation. The dynamics of $K0$ is a stable zero fixed point; see Fig. 8.3a.

KI Sets: Source of Sustained Non-zero Background Activity

A *KI* set is a collection of $K0$ sets, which may consist of either excitatory or inhibitory units. The *KI* set sums activity of many spiking neurons and the summed activity is described in the form of spike density. At the *KI* level, it is important to have the same type of units in the system, so we have no negative feedback. Accordingly, we can talk about excitatory or inhibitory *KI* sets, i.e., KI_e and KI_i, respectively. An interactive excitatory population (KI_e) is modeled with a 2-D layer of $K0$ nodes. Neurons cannot excite themselves, owing to their refractory periods but can only excite each other. The KI_e set activity is described by the solutions to a system of ODEs. The solutions designate state variables that are continuous in time and space and therefore macroscopic [3, 26].

If *KI* has sufficient functional connection density, it will able to maintain a non-zero state of background activity by mutual excitation (or inhibition). The dynamics

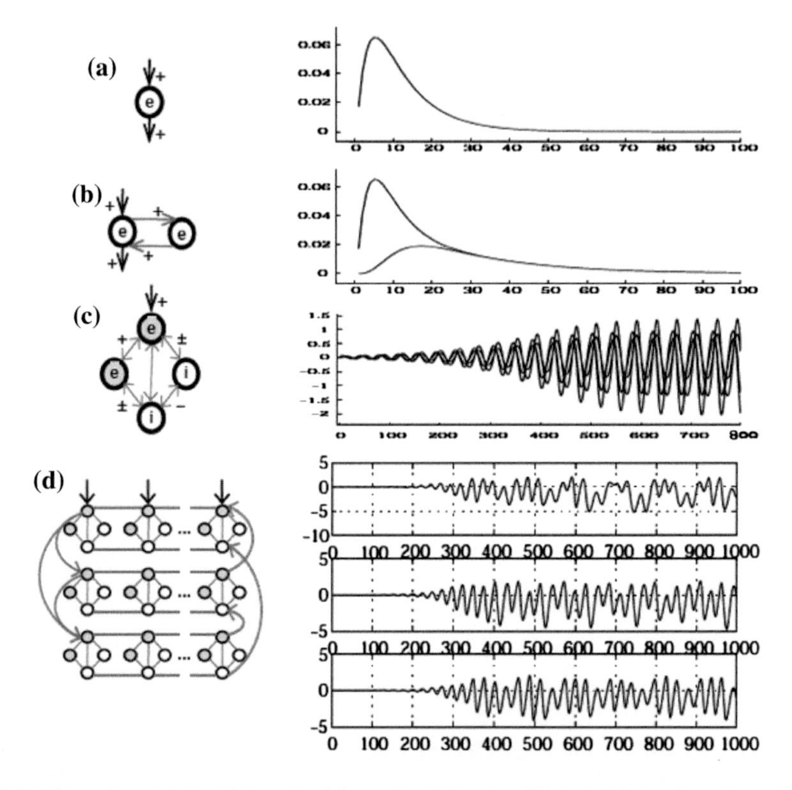

Fig. 8.3 Illustration of the architecture and dynamics of Freeman K sets; **a** K0 set of non-interactive population with stable zero attractor; **b** KI set with mutually excitatory neurons producing non-zero activation level; **c** *KII* with interacting excitatory and inhibitory KI sets producing narrow-band oscillations; **d** *KIII* set with interacting *KII* sets resulting in broad-band oscillations [13]

of a *KI* is described a simple fixed point convergence; see Fig. 8.3b. *KI* typically operates far from thermodynamic equilibrium and the solutions designate macroscopic state variables that are continuous in time and space [3]. The stability of a *KI* set under impulse perturbation is demonstrated using periglomerular cells in the olfactory bulb [4].

Negative Feedback in *KII* Sets to Generate Narrow-Band Oscillations

A major function of the inhibitory neural populations is to selectively enhance sensitivity to reinforced conditional stimuli (CS) under pair-wise excitation. A *KII* set represents a collection of excitatory and inhibitory populations, KI_e and KI_i. A *KII* set incorporates the key role of inhibitory populations and it represents a collection of excitatory and inhibitory cells, KI_e and KI_i. A *KII* set has four types of interactions: excitatory-excitatory, inhibitory-inhibitory, excitatory-inhibitory, and inhibitory-excitatory. Under sustained excitation from a KI_e set but without the equivalent of sensory input, the *KII* set is governed by limit cycle dynamics. With simulated sensory input comprising an order parameter, the *KII* set undergoes a state transition to oscillation at a narrow band frequency in the gamma range. The action of *KII* sets is to selectively enhance narrow band oscillations at frequencies with the beta-gamma range. The prolongation of unit firing is only the prelude to perception; more salient is the enhancement of macroscopic oscillation.

A *KII* set incorporates the key role of inhibitory populations and it represents a collection of excitatory and inhibitory cells, KI_e and KI_i. A *KII* set has four types of interactions: excitatory-excitatory, inhibitory-inhibitory, excitatory-inhibitory, and inhibitory-excitatory. Under sustained excitation from a KI_e set but without the equivalent of sensory input, the *KII* set is governed by limit cycle dynamics; see Fig. 8.3c. With simulated sensory input comprising an order parameter, the *KII* set undergoes a state transition to oscillation at a narrow band frequency in the gamma range. The action of *KII* sets is to selectively enhance narrow band oscillations at frequencies with the beta-gamma range. In aroused states the output of the isolated bulb is periodic oscillation in the gamma range [28], which is modeled with a limit cycle attractor.

Systematic analysis has identified regions of stability and transitions between limit cycle and fixed-point dynamics [7, 8, 26]. The capability of the bulb for limit cycle oscillation in aroused states has been demonstrated for the gamma range after isolation by surgical, pharmacological, or cryogenic blockade [28].

Simulations of olfactory ECoG by the solutions of networks of nonlinear ODE by piece-wise linear approximations show that an increase in the strength of mutual excitation (positive excitatory feedback gain) not only prolonged the KI_e impulse response; it strongly amplified the narrow-band oscillation at the frequency selected by negative feedback between KI_i and *KII* populations. The excitatory bias drove the cortical dynamics into increased negative feedback gain [29]. The change in waveform of the olfactory impulse response with associative learning was simulated by increasing the gain, *kee*, while the change with habituation was simulated with a decrease [30]. The importance of this topological property in network analysis cannot be overemphasized; it is a necessary feature of the mechanism providing the phase transition.

KIII Sets—the Origin of Broad-Band Oscillations

KIII is the result of the interaction of several *KII* sets. If the frequencies of the *KII* sets are incommensurate, *KIII* can general broad-band, chaotic oscillations; see Fig. 8.3d. *KIII* is an example of associative memories. Mesoscopic associative memory systems include synfire chains [31], chaotic recurrent networks [32], the brain-state-in-a-box [33], spin glass models [34], and self-organizing maps [35]. Dynamic *KIII* memories have several advantages as compared to convergent recurrent networks:

1. *KIII* sets produce robust memories based on relatively few learning examples even in noisy environment;
2. The encoding capacity of a network with a given number of nodes is exponentially larger than their convergent counterparts;
3. They can recall the stored data very quickly, just as humans and animals can recognize a learnt pattern within a fraction of a second [13, 19].

KIII sets play an important role in the embodiment of meaning in AM patterns of neural activity shaped by synaptic interactions that have been modified through learning in *KIII* layers.

Sensory systems create spatial patterns of macroscopic neural activity from the microscopic information that was initiated in sensory receptors, and that was preprocessed in the olfactory bulb for smell and the thalamus for all other senses. The negative feedback loops provide the narrow band carrier wave that transports the cortical output and parses it into frames [36]. The positive feedback loops provide the local amplification and attenuation of amplitude of the carrier wave, which results in the spatial texture of amplitude modulation (AM), and which expresses the memory of stimuli that is stored in the modifications of synapses among excitatory cells. The inhibitory positive feedback loops provide the lateral inhibition that sharpens the spatial AM patterns.

8.2 Finite-Size Scaling Theory for Random Graphs

There is considerable progress in the literature in the area of critical behavior in equilibrium systems like Ising models and in non-equilibrium systems, including PCA, coupled map lattices, and directed percolation [37–40]. Various universality classes have been identified for systems exhibiting common critical exponents. It is beyond the scope of this study to elaborate on the achievements on university classes, finite size scaling, and renormalization group theory [41, 42]. For the present treatment it is important to point out that various PCAs display critical behavior interpretable in the context of phase transitions.

In systematic studies of PCA, statistical characteristics of interest include magnetization m, susceptibility χ and correlation length ξ [43]. Finite size scaling theory holds for a large class of lattices [44] and it tells that statistical features calculated with different lattice sizes intersect at a unique point, marking criticality. At criticality, the behavior is independent of the system size, the correlation length is infinity, and signals reach all regions simultaneously, with no delay. This unique point is given by the critical noise p_0.

Binder's method has been developed for characterizing critical behavior in Ising spin glass systems [44]. If the number of active and inactive sites is equal at a given time, the activation density d becomes 0.5. This corresponds to a basal state in magnetic materials with no magnetization. Deviations from the 0.5 level indicate non-zero magnetization $m(t) = |d(t) - 0.5|$. The expected value of magnetization is denoted as $< m >$. In addition to magnetization, important quantities analyzed in Binder's method include the correlation length and susceptibility. The correlation length corresponds to the typical cluster size. The susceptibility describes the variance of the system activation per site. For Ising systems, magnetization, susceptibility,

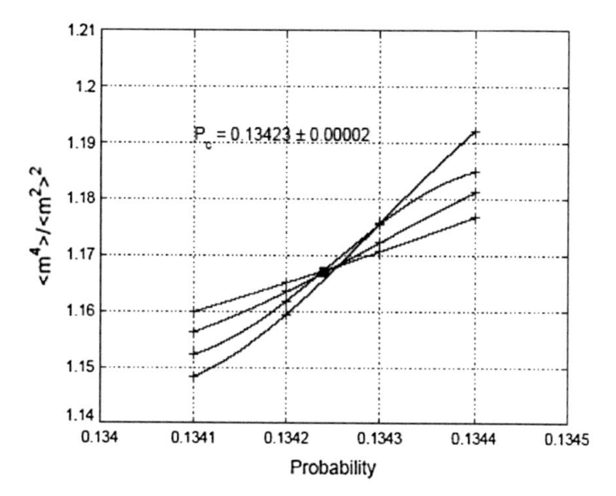

Fig. 8.4 Critical probability estimation using the fourth order cumulants U; the curves correspond to lattice sizes $N = 45, 64, 91$ and 128 [45]

and correlation length satisfy a power law scaling behavior near criticality defined as follows:

$$< m >= (\varepsilon - \varepsilon_c)^{\beta}, \quad \chi \sim |p - p_c|^{-\gamma}, \quad \xi \sim |p - p_c|^{-\nu} \tag{8.3}$$

Finite size scaling theory tells that the fourth order cumulants are expected to intersect each other at a unique point which is independent of the lattice size. The corresponding probability of this unique point is the critical probability ε_c. This scaling property is used in Binder's method to determine the critical parameters, see Fig. 8.4. The fourth order cumulants are defined for a given lattice size N and probability ε as

$$U(N, \varepsilon) =< m^4 > / < m^2 >^2. \tag{8.4}$$

Here β, γ, and ν are critical exponents corresponding to magnetization, susceptibility, and correlation length, respectively. In order to test the consistency of the critical behavior, the hyperscaling relationship $2\beta + \gamma = d\nu$ is evaluated, which holds for Ising systems [44]; here $d = 2$ for the studied two-dimensional lattice geometry. Hyperscaling relationship is considered as a measure of the quality of the estimation of the critical exponents is a given system. The above statistical properties of neuropercolation processes have been evaluated based on computer simulations.

Statistical properties of neuropercolation processes have been evaluated for various rewiring rates ($0 \le q \le 1$) using Monte Carlo simulations. In statistical evaluations, Binder's method of cumulants is used [43, 44]. Simulations have been conducted for n x n lattices of size $n = 64, 92, 128, 256$. Typical calculations require 10^8 or more iterations in order to achieve the desired statistical accuracy of critical exponents.

Table 8.1 shows critical exponents of the 2D lattice as defined in Eq. 8.3. We consider the case of the RCA lattice graph without rewiring, Ising system in 2-D [44], and Toom cellular automaton (TCA) as well [43].

Table 8.1 Critical parameters of local models[a]

Critical parameters	Neuropercolation RCA	Toom CA	Ising 2D
β	0.1308	0.12	0.125
γ	1.8055	1.59	1.75
ν	1.0429	0.85	1
Error	0.02	0.13	0

[a] 2D lattice graph in isolation, without rewiring

Table 8.2 Critical parameters with rewiring

Critical parameters	q = 0 local	q = 6.25 small world	q = 12.5 small world	q = 100 global
β	0.1308	0.3071	0.4217	0.4434
γ	1.8055	1.1920	0.9873	0.9371
ν	1.0429	0.9504	0.9246	0.9026
Error	0.02	0.09	0.09	0.02

The *Error* in the last row indicates the error of the hyperscaling relation evaluated based on the critical exponents for each specific system. As Table 8.1 shows, hyperscaling is satisfied with high accuracy in the studied neuropercolation RCA model. This indicates that the model exhibits behavior close to an Ising model, i.e., it belongs to the *'weak Ising universality class'* [46].

Table 8.1 shows critical exponents of the 2D PCA model as defined by Eq. 8.3 [46, 47]. For comparison, we display critical exponents of PCA models with various level of random rewiring, see Table 8.2. In the case of small world and global systems, the critical exponents deviate significantly from the values obtained for models without rewiring (local models). Still the hyperscaling relationship approximately holds (*Error* ≈ 0). The results confirm the result [43], namely that majority PCA model is not in the Ising universality class. However, the critical exponents exhibit behavior close to the Ising model, i.e., majority PCA belongs to the *weak Ising class*. Therefore, it is justified to talk about critical behavior and phase transitions in PCA models, in spite the fact that no precise proof of criticality exists for these models. This notion of criticality in PCAs is adopted and expanded in neuropercolation models.

References

1. Freeman WJ (2001) How brains make up their minds. Columbia UP, New York
2. Katchalsky Katzir A (1971) Biological flow structures and their relation to chemodiffusional coupling. Neurosci Res Prog Bull 9:397–413
3. Freeman WJ, Erwin H (2008) Freeman K-set. Scholarpedia 3(2):3238

4. Freeman WJ (2000/2006) Neurodynamics. An exploration of mesoscopic brain dynamics. Springer, London. Electronic version, http://sulcus.berkeley.edu/
5. Freeman WJ, Chang H-J, Burke BC, Rose PA, Badler J (1997) Taming chaos: stabilization of aperiodic attractors by noise. IEEE Trans Circuits Syst 44:989–996
6. Kozma R, Freeman WJ (2001) Chaotic resonance: methods and applications for robust classification of noisy and variable patterns. Int J Bifurc Chaos 10:2307–2322
7. Xu D, Principe JC (2004) Dynamical analysis of neural oscillators in an olfactory cortex model. IEEE Trans Neural Netw 15(5):1053–1062
8. Ilin R, Kozma R (2006) Stability of coupled excitatory-inhibitory neural populations application to control multistable systems. Phys Lett A 360:66–83
9. Gutierrez-Galvez A, Gutierrez-Osuna R (2006) Increasing the separability of chemo-sensor array patterns with Hebbian/anti-Hebbian learning. Sens Actuators B: Chem 116(1):29–35
10. Chang HJ, Freeman WJ (1996) Parameter optimization in models of the olfactory neural system. Neural Netw 9(1):1–14
11. Chang HJ, Freeman WJ, Burke BC (1998) Optimization of olfactory model in software to give 1/f power spectra reveals numerical instabilities in solutions governed by aperiodic (chaotic) attractors. Neural Netw 11(3):449–466
12. Kozma R, Freeman WJ (2002) Classification of EEG patterns using nonlinear neurodynamics and chaos. Neurocomputing 44–46:1107–1112
13. Beliaev I, Kozma R (2007) Time series prediction using chaotic neural networks on the CATS benchmark. Neurocomputing 70(13):2426–2439
14. Kozma R (2007) Neuropercolation. Scholarpedia 2(8):1360
15. Harter D, Kozma R (2005) Chaotic neurodynamics for autonomous agents. IEEE Trans Neural Netw 16(3):565–579
16. Harter D, Kozma R (2006) Aperiodic dynamics and the self-organization of cognitive maps in autonomous agents. Int J Intell Syst 21(9):955–972
17. Kozma R, Freeman WJ (2003) Basic principles of the KIV model and its application to the navigation problem. J Integr Neurosci 2(1):125–146
18. Kozma R, Freeman WJ, Erdi P (2003) The KIV model D nonlinear spatio-temporal dynamics of the primordial vertebrate forebrain. Neurocomputing 52–54:819–826
19. Kozma R, Freeman WJ (2009) The KIV model of intentional dynamics and decision making. Neural Netw 22(3):277–285
20. Huntsberger T, Tunstel E, Aghazarian H, Kozma R (2006) Onboard learning strategies for planetary surface rovers. Intell Space Robot 403–422
21. Kozma R, Huntsberger T, Aghazarian H, Tunstel E, Ilin R, Freeman WJ (2008) Intentional control for planetary rover SRR2K. Adv Robot 21(8):1109–1127
22. James W (1893) The principles of psychology. H. Holt, New York
23. Kandel ER, Schwartz JH, Jessell TM (2000) Principles of neuroscience. McGraw Hill, New York
24. Dayan P, Abbott LF (2001) Theoretical neuroscience. MIT Press, Cambridge
25. Izhikevich EM (2006) Dynamical systems in neuroscience–the geometry of excitability and bursting. MIT Press, Cambridge
26. Freeman WJ (1975/2004) Mass action in the nervous system. Academic, New York. Electronic version 2004. http://sulcus.berkeley.edu/MANSWWW/MANSWWW.html
27. Freeman WJ (1967) Analysis of function of cerebral cortex by use of control systems theory. Logist Rev 3:5–40
28. Gray CM, Skinner JE (1988) Centrifugal regulation of neuronal activity in the olfactory bulb of the waking rabbit as revealed by reversible cryogenic blockade. Exp Brain Res 69:378–86
29. Freeman WJ (1979) Nonlinear dynamics of paleocortex manifested in the olfactory EEG. Biol Cybern 35:21–37
30. Emery JD, Freeman WJ (1969) Pattern analysis of cortical evoked potential parameters during attention changes. Physiol Behav 4:67–77
31. Abeles M (1991) Corticonics: neural circuits of the cerebral cortex. Cambridge UP, New York
32. Aihara K, Takabe T, Toyoda M (1990) Chaotic neural network. Phys Lett A 144:333–340

33. Anderson JA, Silverstein JW, Ritz SR, Jones RS (1977) Distinctive features, categorical perception, and probability learning: some applications of a neural model. Psychol Rev 84:413–451
34. Hopfield JJ (1982) Neuronal networks and physical systems with emergent collective computational abilities. Proc Natl Acad Sci USA 81:3058–3092
35. Kohonen T (2001) Self-organizing maps. Springer, Berlin
36. Freeman WJ (2005) Origin, structure, and role of background EEG activity. Part 3. Neural frame classification. Clin Neurophysiol 116(5):1118–1129
37. Makowiec D, Gnacinski P (2002) Universality class of probabilistic cellular automata. Cellular automata. Springer, Berlin, pp 104–113
38. Turova TS (2012) The emergence of connectivity in neuronal networks: from bootstrap percolation to auto-associative memory. Brain Res 1434:277–284
39. Regnault D (2013) Proof of a phase transition in probabilistic cellular automata. Developments in language theory. Springer, Berlin, pp 433–444
40. Turova T, Vallier T (2015) Bootstrap percolation on a graph with random and local connections. arXiv preprint arXiv:1502.01490
41. Kadanoff LP, Ceva H (1971) Determination of an operator algebra for a two-dimensional Ising model. Phys Rev B3:3918
42. Odor G (2004) Universality classes in nonequilibrium lattice systems. Rev Mod Phys 76:663–724
43. Makowiec D (1999) Stationary states for Toom cellular automata in simulations. Phys Rev E 60:3787–3796
44. Binder K (1981) Finite scale scaling analysis of Ising model block distribution function. Z Phys B 43:119–140
45. Puljic M, Kozma R (2005) Activation clustering in neural and social networks. Complexity 10(4):42–50
46. Kozma R, Puljic M, Balister P, Bollobas B, Freeman WJ (2005) Phase transitions in the neuropercolation model of neural populations with mixed local and non-local interactions. Biol Cybern 92:367–379
47. Puljic M, Kozma R (2008) Narrow-band oscillations in probabilistic cellular automata. Phys Rev E 78026214

Chapter 9
Supplement II: Signal Processing Tools

9.1 Description of ECoG and EEG Signals

Brain waves are commonly treated as if they were the sum of the outputs of a set of neural oscillators, each of which has a constant frequency and variable amplitude. This treatment is based on the assumption that brain dynamics is linear and time-invariant, which is clearly not the case. The advantage conveyed by this assumption is the ease with which linear analysis can be applied to brain waves using, e.g., Fast Fourier Transform (FFT). The disadvantage is the inability of the linear analysis to capture and display the transient dynamics, including nonlinear state transitions by which brains operate. The decomposition into amplitude and phase functions is by the Fourier transform giving high resolution of frequency in stationary signals. If the signals are nonstationary, non-Gaussian, and significantly deviate from linear behavior, alternative approaches can be justified. In particular, the Hilbert Transform (HT) may be beneficial by giving high temporal resolution of signals undergoing phase and frequency modulation. For a review of the widely employed signal analysis tools for brain state monitoring, see [1, 2].

Before describing nonlinear signal processing tools, first we briefly address linear approaches [3]. Common practice is to assume linearity and stationarity in the dynamics of their sources, and to decompose the signals into derivative state variables by use of, e.g., principal component analysis (PCA), independent component analysis (ICA), auto-regression (AR-MA). Prominent among these transforms is the fast Fourier transform (FFT), which decomposes each raw signal into a family of state variables having fixed frequencies and amplitudes (periodic components). Closely related are wavelets that are used for linear decomposition of signals varying in amplitude. The temporal resolution of wavelets and the FFT is bounded by the Nyquist criterion: the digitizing rate must be at least twice and preferably three times the highest component frequency. The duration of segments for decomposition must exceed at least one cycle of the lowest component frequency [1, 2, 4].

R. Kozma and W.J. Freeman, *Cognitive Phase Transitions in the Cerebral Cortex – Enhancing the Neuron Doctrine by Modeling Neural Fields*,
Studies in Systems, Decision and Control 39, DOI 10.1007/978-3-319-24406-8_9

For example, consider the ECoG recorded from 8×8 high-density arrays [5], which reveal high correlation among the $N = 64$ signals; typically 95 % of the variance in PCA is incorporated in the first component. Commonly the correlation is attributed to activity at the reference electrode in referential recording and to volume conduction. These factors do not account for the high correlation, as shown by the 2-fold differences in amplitude and phase of the shared carrier waveform often seen between signals from adjacent electrodes, and as proved by calculation of the dendritic point spread function of the cortical generators [6]. The 64 values of amplitude have behavioral correlates; the 64 values of phase do not, but they reveal the manner and mechanism by which cortical dynamics makes patterns. FFT analysis reveals important relationships between the amplitude patterns and cognitive activities, such as classification and decision making [7, 8], however, nonlinear methods provide additional insights, as described next.

9.2 Hilbert Transform and Analytical Signal Concept for Pattern Analysis

9.2.1 Basic Concepts of Analytic Signals

An alternative transform to FFT is the Hilbert transform [3, 9], which when applied to a brain wave recording in effect calculates the rate of change in the amplitude at each time step of the digitized signal. This operation effectively re-expresses an oscillation as a vector that rotates counterclockwise in the complex plane. The amplitude is expressed by the length of the vector, $A(t)$, and the rate of change is expressed by the angular velocity of the rate of rotation of the vector about the origin of the complex plane. The rate of rotation is expressed as a rate of change in phase in degrees/second, radians/second (rad/s), or cycles/second (Hz). The immediate advantage is that the Hilbert transform decomposes a brain wave into an analytic amplitude, $A(t)$, and an analytic phase, $P(t)$. The change in phase in rad with each time step divided by the digitizing interval in s approximates an instantaneous frequency that can vary, unlike the frequencies that are extracted by Fourier decomposition. The disadvantage is that the Hilbert transform is very sensitive to noise of many kinds; it only works well after band pass filtering of a brain wave. Criteria for optimal band pass filtering have been described elsewhere [10–12].

The application of the Hilbert transform to each intracranial recording from an array of microelectrodes is a multi-step procedure [6]. First, a high pass filter extracts the mean signal amplitude, and a low pass filter extracts the LFP from the microelectrode recordings. Second, the low pass data are down-sampled to 500 Hz and normalized to zero mean for every channel and unit standard deviation (SD) for all channels. Third, the demeaned, normalized LFP are band pass filtered in the classic empirical ranges: theta (3–7 Hz), alpha (7–12 Hz), beta (12–30 Hz), gamma (30–60 Hz), and high gamma (60–200 Hz), and the n channels in each pass band are segmented to save the data from each trial. Fourth, the Hilbert transform is applied to get the analytic signal, $V_j(t)$ in radians by using the Hilbert transform [13–15]. The signal of each channel is denoted as $v_j(t)$, $j = 1, \ldots, N$. The real-valued time series $v_j(t)$ obtained at a trial is transformed to a complex time series $V_j(t)$, with a real part $v_j(t)$ and imaginary part, $u_j(t)$:

$$V_j(t) = v_j(t) + iu_j(t), \quad j = 1, \ldots, N. \tag{9.1}$$

Here, the real part is the original time series (Fig. 9.1 blue curve), while the imaginary part is derived from the Hilbert transform of $v_j(t)$, denoted as $u_j(t)$ (Fig. 9.1 red curve):

$$u_j(t) = \frac{1}{\pi} P.V. \int_{-\infty}^{+\infty} \frac{v_j(t')}{t - t'} dt', \tag{9.2}$$

where P.V. signifies the Cauchy Principal Value. At each digitizing step the EEG yields a point in polar plot in the complex plane. Sequences of steps give a trajectory of the tip of a vector rotating counterclockwise in the complex plane with elapsed time. The vector length at each digitizing step, t, is the analytic amplitude (the blue curve in Fig. 9.1c),

$$A_j(t) = \sqrt{\left(v_j^2(t) + u_j^2(t)\right)}, \tag{9.3}$$

and the arctangent of the angle of the vector with respect to the real axis is the analytic phase (blue sawtooth in Fig. 9.1d),

$$P_j(t) = atan\left(\frac{u_j(t)}{v_j(t)}\right), \quad j = 1, \ldots, N. \tag{9.4}$$

9.2.2 Amplitude Modulation (AM) Patterns

The mean of the square of amplitudes $A_j^2(t)$ over N channels gives the mean power $A^2(t)$. Amplitudes $A_j(t)$ are frequently normalized by the ensemble average to yield the normalized feature vector at each time step. The feature vector specifies the normalized spatial pattern formed in the pass band by the signals from the N channels. It designates a point in N-space that is occupied by the tip of the feature vector as it describes a trajectory through infinite brain state space that is projected into N-space by measurement. The feature vector provides a measure of the order parameter of the ensemble of cortical neurons that is under observation [10].

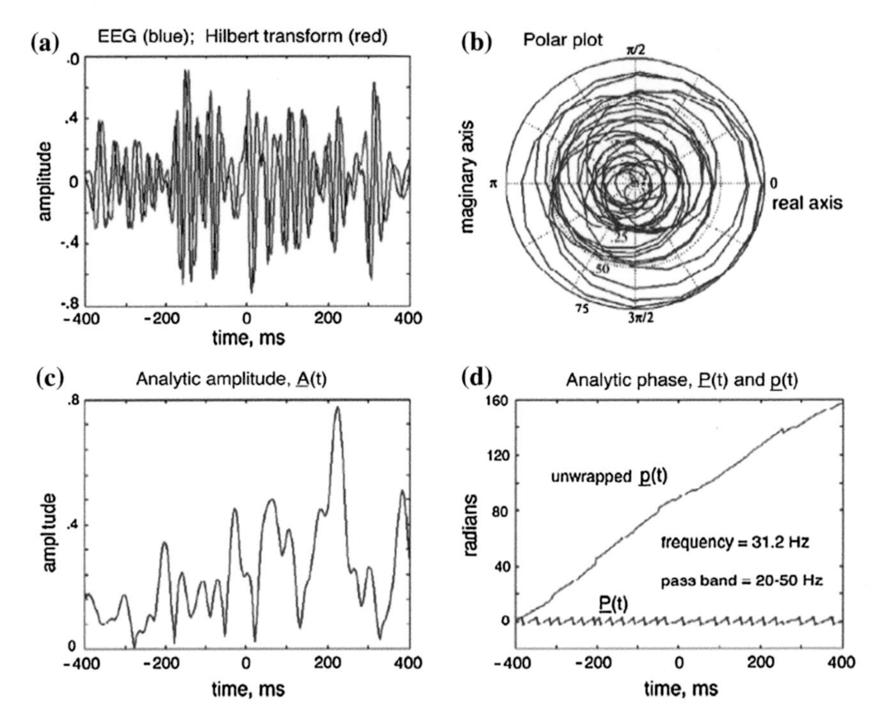

Fig. 9.1 Application of the Hilbert Transform to 64-channel EEG in an 800 ms epoch of visual cortical EEG centered at delivery of a CS at 0 ms. **a** A representative segment on a typical channel was selected from visual cortical beta-gamma EEG after band pass filtering (20–50 Hz). The *blue curve* shows the spatial ensemble average of the real part representing the excitatory neuronal output ($v(t)$. The *red curve* shows the imaginary part representing the inhibitory neuronal output $u(t)$. **b** The real part of the analytic signal (abscissa) is plotted against the imaginary part (ordinate) as a vector. Time is implicit in counterclockwise rotation of the vector tip starting from the asterix just to the right of the origin where the axes cross. **c** The *blue curve* shows the average analytic amplitude, $A(t)$, in Eq. 9.3 giving the length of the vector). **d** The *blue sawtooth curve* shows the average analytic phase, $P(t)$, given by Eq. 9.4. The *red curve* shows the average unwrapped phase, $p(t)$. The analytic frequency is taken from the slope in rad/s. The deviations from the average slope show *phase slip* which is due to repeated state transitions. A reinforced conditioned stimulus (CS+, full field dim light flash) was delivered at 0 ms; from [10]

9.2.3 Frequency Modulation (PM): Temporal Resolution of Frequency

The decomposition into amplitude and phase functions is by the Fourier transform giving high resolution of frequency in stationary signals, and by the Hilbert transform giving high temporal resolution of signals undergoing phase and frequency modulation. The 64 values of amplitude have behavioral correlates; the 64 values of phase do not, but they reveal the manner and mechanism by which cortical dynamics makes patterns [2, 11].

References

1. Nunez PL, Srinivasan R (2006) Electric fields of the brain: the neurophysics of EEG, 2nd edn. Oxford University Press, New York
2. Freeman WJ, Quian Quiroga R (2013) Imaging brain function with EEG: advanced temporal and spatial analysis of electroencephalographic and electrocorticographic signals. Springer, New York
3. Freeman WJ (2007) Hilbert transform for brain waves. Scholarpedia 2(1):1338
4. Berthouze L, James LM, Farmer SF (2010) Human EEG shows long-range temporal correlations of oscillation amplitude in Theta, Alpha and Beta bands across a wide age range. Clin Neurophysiol 121(8):1187–1197
5. Barrie JM, Freeman WJ, Lenhart M (1996) Modulation by discriminative training of spatial patterns of gamma EEG amplitude and phase in neocortex of rabbits. J Neurophysiol 76:520–539
6. Freeman WJ, Holmes MD, West GA, Vanhatalo S (2006) Fine spatiotemporal structure of phase in human intracranial EEG. Clin Neurophysiol 117:1228–1243
7. Freeman WJ (2000/2006) Neurodynamics. An exploration of mesoscopic brain dynamics. Springer, London Electronic version http://sulcus.berkeley.edu/
8. Freeman WJ, Barrie JM (2000) Analysis of spatial patterns of phase in neocortical gamma EEGs in rabbit. J Neurophysiol 84(3):1266–1278
9. Stam CJ (2005) Nonlinear dynamical analysis of EEG and MEG: review of an emerging field. Clin Neurophysiol 116(10):2266–2301
10. Freeman WJ (2004) Origin, structure, and role of background EEG activity. Part 1. Analytic amplitude. Clin Neurophysiol 115:2077–2088
11. Freeman WJ (2004) Origin, structure, and role of background EEG activity. Part 2. Analytic phase. Clin Neurophysiol 115:2089–2107
12. Freeman WJ (2005) Origin, structure, and role of background EEG activity. Part 3. Neural frame classification. Clin Neurophysiol 116(5):1118–1129
13. Barlow JS (1993) The electroencephalogram: its patterns and origins. MIT Press, Cambridge
14. Pikovsky A, Rosenblum M, Kurths J (2001) Synchronization a universal concept in non-linear sciences. Cambridge University Press, Cambridge
15. Freeman WJ, Burke BC, Holmes MD (2003) Aperiodic phase re-setting in scalp EEG of beta-gamma oscillations by state transitions at alpha-theta rates. Hum Brain Mapp 19(4):248–272

Chapter 10
Supplement III: Neuroanatomy Considerations

10.1 Structural Connectivities: Emergence of Neocortex from Allocortex

The laminar neuropil in the phylogenetically older parts of the cerebrum is called allocortex [1]. Examples include the prepyriform cortex (paleocortex), the hippocampus (archicortex) and parts of the perirhinal and entorhinal cortex (mesocortex) [2]; the olfactory bulb is here included as allocortex owing to its similarity to the others in topology and phylogenetic derivation, though not all anatomists accept this taxonomy. Generically allocortex has three layers with differing subdivisions specific to each area. Layer I also called marginal lies under the bounding pial membrane and has input axons and the dendritic trees on which they synapse. Layer II has the cell bodies, often with triangular shapes giving the name pyramidal cells (mitral cells in the bulb). Layer III has output axons with recurrent side branches called collateral branches that synapse on interneurons called stellate cells (internal granule cells in the bulb) but mainly on other pyramidal cells.

In contrast to allocortex with three layers, neocortex (also called isocortex) is described as having six layers on the basis of selective staining of cell nuclei, though with many variations and exceptions [3, 4]. Comparative neuroanatomists have shown how neocortex (found only in mammals) has evolved from reptilian allocortex by intrusion into allocortical Layer II (Fig. 10.1) of neurons migrating from the floor of the basolateral forebrain [5] or dorsal pallium [6]. Neocortical cells are guided by a transitory subplate [7, 8] that enables newer cells to leapfrog over pioneer cells by inside-out progression and form the fine-grain radial bundles of parallel fibers that are oriented perpendicular to the pia in neocortical columns and hypercolumns. The cytoarchitectures and connectivity patterns in Layers II–IV vary greatly between local areas. These palisades of axons provide topographic mapping of sensory input and motor output to and from neocortex. The local areas that reveal specializations are labeled by Brodmann. These specializations support the variety of functions that are performed by local networks in the many sensory, associational,

R. Kozma and W.J. Freeman, *Cognitive Phase Transitions in the Cerebral Cortex – Enhancing the Neuron Doctrine by Modeling Neural Fields*,
Studies in Systems, Decision and Control 39, DOI 10.1007/978-3-319-24406-8_10

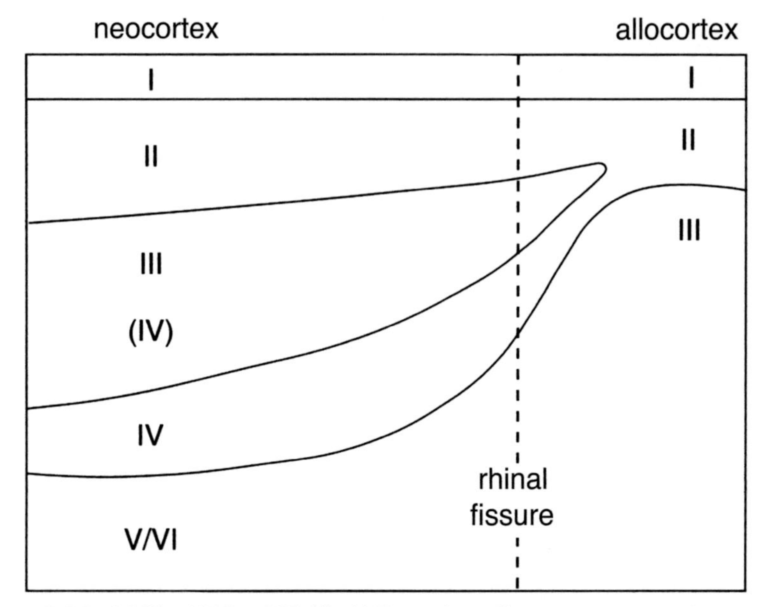

Fig. 10.1 Schematic summary of the transition from allocortex to neocortex by intrusion of cells into Layer II of allocortex through transitional mesocortex around the landmark separating neocortex from allocortex, the rhinal fissure. Adapted from [10], p. 237

and motor areas of neocortex. The individuation is enhanced by close dependence on topographically organized specific thalamic inputs by axons forming synapses in Layer IV. The neocortical marginal Layer I continues to receive input axons from nonspecific thalamic nuclei and from allocortex [9].

Whereas the several areas of allocortex are separated by discontinuities in the neuropil yet are homogeneous within, neocortex is a continuous sheet of neuropil in each cerebral hemisphere. Its specializations are contained within Layers II–IV, while Layers I, V and VI unify them. The deep pyramidal cells in Layers V and VI have long basal dendrites that receive connections from neurons broadly situated in Layers II–IV, and their apical dendrites extend and branch into Layer I. The deep pyramidal cells send long output axons to other areas of cortex, to the basal ganglia and brainstem, and by widely radiating collaterals into the overlying layers. They constitute the broadly interconnected embedding neuropil through which the microscopic activity patterns in the clusters and networks of Layers II–IV might be synchronized into mesoscopic signals, and the results disseminated and integrated at the macroscopic level of modules and lobes in the second of the of the *very few short steps*.

Most analyses of anatomical data have emphasized point-to-point networks, for example, the meta-study [11] of connectional architecture in visual cortex, and the cortico-subcortical modules of [12]. These linkages are very important for the organization of the order parameters of mesoscopic and macroscopic patterns of amplitude modulation (AM) of the shared carrier frequency. However, they do not address the connectivity required for the long-range coherence and rapid state transitions revealed by carrier waves. Braitenberg and Schuz wrote: "A recent hypothesis by Miller [13], based on differences in both connectivity patterns and spontaneous activity between upper and lower layers, assigns to the upper layers the role of a neuronal "library", storing most of the information encoded by assemblies, while the lower layers are assumed to catalyze the process of assembly formation" ([14], p. 150).

Long cortico-cortical connections are not randomly distributed within the territory that they innervate. Instead they are distributed in patches with high local connection density, with intervening regions having few connections [16]. Kaas wrote: "Generalizing from cats and monkeys it appears that the evolutionary advance in brain organization is marked by increases in the numbers of unimodal sensory fields, not by increases in multimodal association cortex as traditionally thought." [17] (p. 147). Therefore Layers II–IV can be conceived to provide highly textured modules, while the embedding neuropil in Layers I, V and VI provides not only the connectivity needed for rapid integration of macroscopic modular activity in the encompassed modules but also the uniformity and unity of neocortical anatomy that has often been noted by neuroanatomists.

10.2 Constancy of Properties of Neocortex Across Species

There is agreement among anatomists [4, 15, 18] on the remarkable invariance across areas of neocortex and species of mammals of the number of neurons/mm^2 being close to 3×10^5/mm^2. The logarithm of thickness of both allocortex and neocortex varies across species in accordance with the logarithm of body size (Fig. 10.2a). The neural density in the human neocortex with 2.6 mm average thickness is 10^5/mm^3. Apart from the scaling factor of body size that holds for allocortex, the only way by which neocortex increases in size with evolution is by increase in surface area by adding neurons at fixed density per unit surface area at the pia. The extra degree of freedom allows neocortex to increase in area disproportionately to the increase in volume relating to body size and therefore to gyrification by wrinkling of the surface, which expands the range of neocortical surface area nearly 2 orders of magnitude above that of allocortex (Fig. 10.2b), giving evidence for self-similar connectivity in support of scale-free dynamics.

Remarkably, from the limited EEG data available across exotic species, the manifestations of neocortical temporal dynamics appear to be similar across four orders of magnitude from the mouse [19] to the whale [20]. The phylogenetic emergence of neocortex clearly opened a new dimension in the evolution of vertebrates that enabled mammals to out-perform other species in direct competition, as dolphins out-swim sharks, and bats out-fly birds. Further data are needed to measure the distributions of axon lengths and construct accurate functions that are needed to specify connectivities; to measure phase velocities in state transitions that constrain the diameters of functional domains; and to evaluate the exponents of the power spectra and

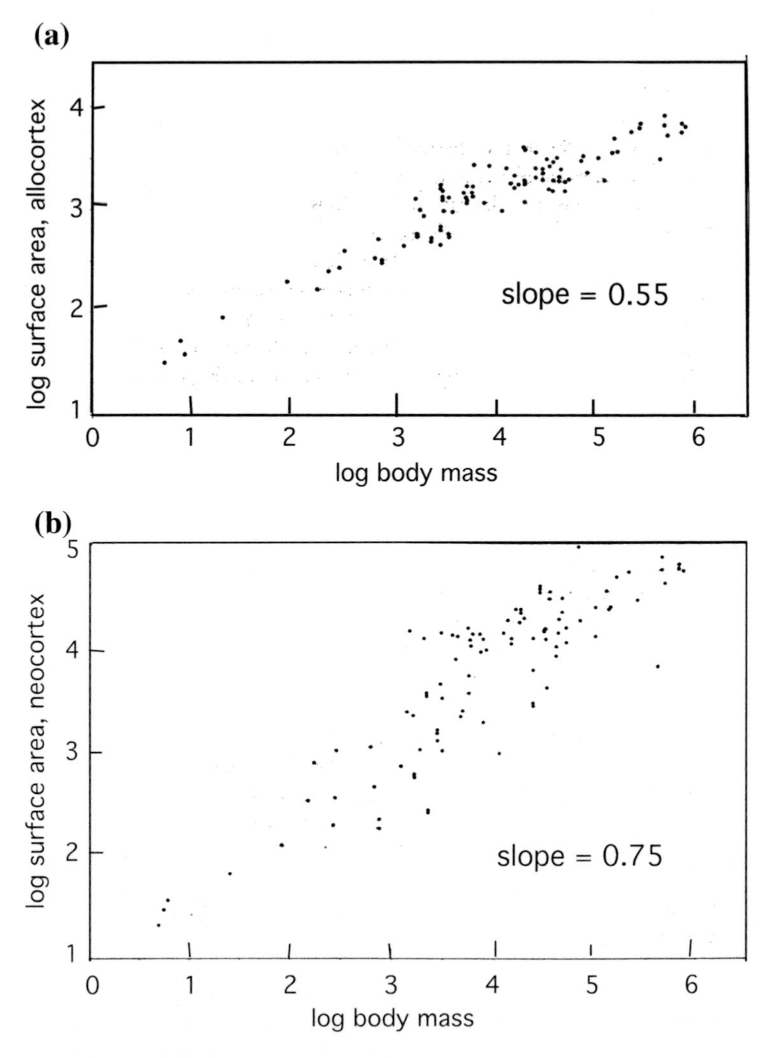

Fig. 10.2 Illustration of the variation of the log area versus body weight across species. **a** Allocortex. **b** Neocortex; from [15], Fig. 110, p. 241

other power-law distributions such as the durations and diameters of phase cones across species, in order to document fully the parameters of scale-free dynamics in macroscopic neural activity.

10.3 Discussion of Scale-Free Structural and Functional Networks

The distributions of cortical connection lengths appearing to be exponential as commonly inferred may actually be power-law, because of three experimental limitations on determinations of distributions of axonal lengths:

First, for short distances the observations using light microscopy omit most unmyelinated axons, which in electron micrographs substantially outnumber the myelinated axons (Fig. 10.3). Second, gap junctions at ultra-short distances may complement synaptic transmission. Third, the observations of axon lengths in Golgi preparations are made in tissue sections of limited thickness (e.g., 300 μ, Fig. 10.3b) in small mammals (here mouse) [14]. Continuity of axons is very difficult to follow in serial sections, leading to a deficit in the numbers of long connections. Also, the longest terminal axonal branches may be smaller than the wavelengths used in optical microcopy, foreshortening the distributions.

Measurements of the diameters of nuclei in pyramidal cells in neocortex [15] as a function of distance from the surface show power-law distributions with different slopes for superficial and deep layers, see Fig. 10.4. The smaller nuclei indicate more local connectivity, while the larger nuclei indicate long connections. Schroeder has shown [22] that convolving an exponential distribution with a hyperbolic distribution of size gives a power-law distribution.

Considering the well-documented self-similarity of many axonal and dendrite trees [23, 24] and pending further anatomical investigation of control mechanisms in large-scale embryological growth, it appears safe to conclude that the distributions of cortical structural connectivity are power-law. This conclusion has been reached by

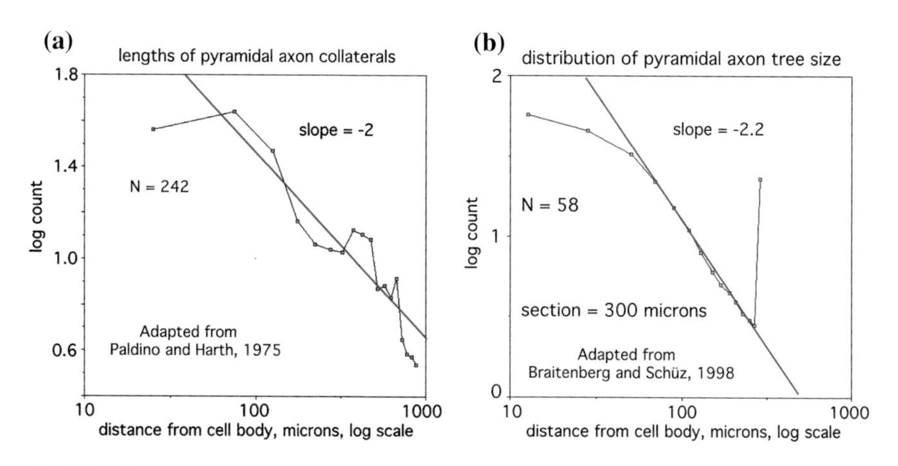

Fig. 10.3 Distributions of measurements of lengths of axons that were made in histological sections of Golgi preparations; **a** length of pyramidal axon collaterals; **b** size of pyramidal axon trees. The data were re-plotted in log–log coordinates [14, 21]

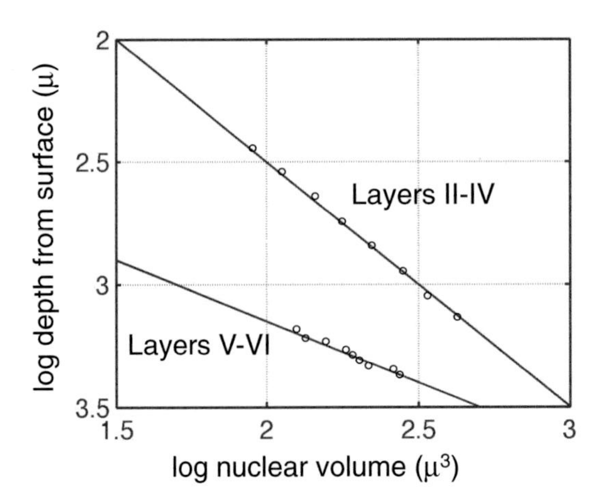

Fig. 10.4 The log volume in cubic microns of pyramidal cell nuclei varies with log depth in microns. The slope $= -1.0$ $(1/f^{\alpha}, \alpha = 1)$ in Layers II–IV; slope $= -0.5$ $(\alpha = 0.5)$ in Layers V and VI. Adapted from Book [15], (p. 104)

several groups [10, 25, 26], and it is important for interpreting physiological evidence regarding functional connectivity. If neocortical connectivity can be demonstrated to conform to a structural scale-free network, and if that supports functional scale-free networks, then for every cognitive function that can be adequately controlled, one or more hubs should be observable at which connection and activity densities are maximal. The foci of maximal activity invariant with behavioral measures revealed by EEG [27], fMRI, e.g., [28], and unit recording [29–31] may express hubs in non-local fields, and they may be decisive in perception.

References

1. Mountcastle VB (ed) (1974) Medical physiology, 13th edn. C V Mosby, St Louis
2. Maclean PD (1969) The triune brain. Plenum, New York
3. Brodmann K (1909) Vergleichende Lokalizationslehre der Grosshirnrinde. Barth, Leipzig
4. Sholl DA (1956) The organization of the cerebral cortex. Methuen-Wiley, London
5. Karten HJ (1997) Evolutionary developmental biology meets the brain: the origins of mammalian cortex. Proc Natl Acad Sci 94:2800–2804
6. Aboitiz F, Montiel J, Morales D, Concha M (2002) Evolutionary divergence of the reptilian and the mammalian brains: considerations on connectivity and development. Brain Res Rev 39:141–153
7. Shatz CJ, Chun JJM, Luskin MB (1988) The role of the subplate in the development of the telencephalon. In: Jones EG, Peters A (eds) The cerebral cortex. The development of the cerebral cortex. Vol III. Plenum, New York, pp 35–58
8. Allendoerfer KL, Shatz CJ (1994) The subplate, a transient neocortical structure: its role in the development of connections between thalamus and cortex. Annu Rev Neurosci 17:185–218
9. Nauta JH (1954) Terminal distributions of some afferent fiber systems in the cerebral cortex. Anat Rec 118:333–346

10. Miller R, Maitra R (2002) Laminar continuity between neo- and meso-cortex: the hypothesis of the added laminae in neocortex. Chap. 11. In: Schuz A, Miller R (eds) Cortical areas: unity and diversity. Taylor and Francis, New York, pp 219–242
11. Fellemin DJ, Van Essen DC (1991) Distributed hierarchical processing in the primate cerebral cortex. Cereb Cortex 1:1–47
12. Houk JC (2005) Agents of the mind. Biol Cybern 92(6):427–437
13. Miller R (1996) Neural assemblies and laminar interactions in the cerebral cortex. Biol Cybern 75:253–261
14. Braitenberg V, Schuz A (1998) Cortex: statistics and geometry of neuronal connectivity, 2nd edn. Springer, Berlin
15. Bok ST (1959) Histonomy of the cerebral cortex. Elsevier, Amsterdam
16. Malach R (1994) Cortical columns as devices for maximizing neuronal diversity. TINS 17: 101–104
17. Kaas JH (1987) The organization of neocortex in mammals: implications for theories of brain function. Annu Rev Psychol 38:129–152
18. Schuz A, Miller R (eds) (2002) Cortical measures: unity and diversity. Taylor and Francis, New York
19. Franken P, Malafosse A, Tafti M (1998) Genetic variation in EEG activity during sleep in inbred mice. Am J Physiol 275 RICP 44:R1127–1137
20. Lyamin OI, Mukhametov LM, Siegel JM, Nazarenko EA, Polyakova IG, Shpak OV (2002) Unihemispheric slow wave sleep and the state of the eyes in a white whale. Behav Brain Res 129:125–129
21. Paldino A, Harth E (1977) A computerized study of Golgi-impregnated axons in rat visual cortex. In: Lindsay RD (ed) Computer analysis of neuronal structures. Plenum, New York, pp 189–207
22. Schroeder M (1991) Fractals, chaos power laws. W.H. Freeman, New York
23. Uylings HBM, Van Pelt J (2002) Measures for quantifying dendritic arborizations. Netw Comput Neural Syst 13:397–414
24. Jelinek HJ, Elston GN (2003) Dendritic branching of pyramidal cells in the visual cortex of the nocturnal owl monkey: a fractal analysis. Fractals 11(4):1–5
25. Linkenkaer-Hansen K, Nikouline VM, Palva JM, Iimoniemi RJ (2001) Long-range temporal correlations and scaling behavior in human brain oscillations. J Neurosci 15:1370–1377
26. Hwa RC, Ferree T (2002) Scaling properties of fluctuations in the human electroencephalogram. Phys Rev E 66:021901
27. Basar E (2005) Memory as the whole brain work a large-scale model based on oscillations in super-synergy. Int J Psychophysiol 58:199–226
28. Buxton RB (2001) Introduction to functional magnetic resonance imaging: principles and techniques. Cambridge University Press, Cambridge
29. Abeles M (1991) Corticonics: neural circuits of the cerebral cortex. Cambridge University Press, New York
30. Singer W, Gray CM (1995) Visual feature integration and the temporal correlation hypothesis. Annu Rev Neurosci 18:555–586
31. Quian Quiroga R, Reddy L, Kreiman G, Koch C, Fried I (2005) Invariant visual representation by single-neurons in the human brain. Nature 435:1102–1107

Part III
Commentaries on Neuroscience Experiments at Cell and Population Levels

Chapter 11
Commentary by B. Baars

How Does the Cortex Know? A Walk Through Freeman Neurodynamics

Bernard J. Baars

Abstract Freeman Neurodynamics may be complementary to conventional analysis of direct cortical recording. (ECoG) Freeman and Kozma have discovered nonlinear dynamical phenomena in cortex, involving the dendritic feltwork of Layer I. After Hilbert analysis, a strip of electrodes with sufficient spatial density reveals intermittent phase equilibria interrupted by $\sim$10 Hz phase collapses. This activity propagates via phase dispersion. Kozma and Freeman, Advances in cognitive neurodynamics, 2015, [10] propose that "the rapid propagation of phase dispersion over the hemisphere is the manifestation of the cognitive broadcast as described in Baars' global workspace theory." Sensory contents might be expressed in "spatial patterns of amplitude modulation of beta and gamma waves." This view emerges from Freeman K sets, including recent modeling using the concept of neuropercolation in sets of cellular automata. It also implies scale-free dynamics and inverse power-law distributions of cortical connectivity and EEG power. Yet what is it that is being propagated? In the conventional view, conscious visual percepts emerge from more than 40 visuotopical activity arrays linked bidirectionally, so that *adaptive resonance* enables a problem-solving trajectory terminating in a coherent visual gestalt via winner-take-all competition. New findings indicate that such visual gestalts emerge in at high levels of the visual hierarchy, whence they are accurately propagated to the prefrontal lobe, consistent with global workspace theory Panagiotaropoulos et al., Neuron 74(5):924–935, 2012, [12] and Baars et al., Front Psychol 4:200, 2013, [2]. Conventional ECoG analysis may reflect the *cognitive* aspect of global binding and broadcasting, while Freeman Neurodynamics may convey intentional information—biological goals and conditioned stimuli. In Franklin's terms, the dendritic neuropil may shape action selection, while axonal resonance may support perceptual contents. One possibility is that cortex evolved a variety of Sparse Distributed Memory (SDM) Kanerva, Sparse distributed memory, 2003, [7] and Snaider and Franklin, Cogn. Comput. 6(3):510–527, 2014, [13], a very efficient, non-local way to pack information into high-dimensional bit vectors.

B.J. Baars
The Neurosciences Institute, La Jolla, CA, USA

B.J. Baars
4705 Frazee Rd, Apt 102, Oceanside, CA 92057, USA

R. Kozma and W.J. Freeman, *Cognitive Phase Transitions in the Cerebral Cortex – Enhancing the Neuron Doctrine by Modeling Neural Fields*,
Studies in Systems, Decision and Control 39, DOI 10.1007/978-3-319-24406-8_11

This suggests that some brain region acts as a minimal retrieval vector for episodic learning. The hippocampus and the claustrum may be candidates for such an SDM vector substrate.

11.1 Introduction

WJ Freeman and Robert Kozma have developed a strikingly novel approach to mass action in the brain, especially the horizontal dendritic neuropil of cortex. This paper considers selected hypotheses from Kozma and Freeman (in press) from the viewpoint of Dynamic Global Workspace Theory, a rigorous effort to account for conscious (reportable) brain events [2].

11.1.1 Does the Cortex "know" or "intend"?

Conventional textbook accounts of cortex take a cognitive approach, claiming that "knowing" is a key function of cortex. WJ Freeman [6] and Stan Franklin [5] adopt more of a pragmatist point of view, in which the primary function of cortex is intentional set and action selection.

Conventional intracranial recordings in the macaque reveal more than forty visuotopical arrays, which are linked bidirectionally, so that resonant activity emerges, converging via a winner-take-all process to coherent visual gestalts. Recent findings show that high-level visual gestalts emerge in MTL (medial temporal lobe), from where gestalt is accurately propagated to the prefrontal lobe, as predicted by global workspace theory [1, 2, 12].

Fourier analytical methods therefore seem to support a cognitive epistemology, with active spatial arrays showing increasingly integrative levels of analysis of sensory input, converging on a conscious gestalt followed by accurate report.

Freeman Neurodynamics originally used Pavlovian conditioning in rabbits, in which neutral stimuli become learned predictors for biological stimuli. This is consistent with Stan Franklin's approach, which claims that *action selection* is the essential function of animal and robot brains [5].

Can the two approaches give a single, coherent understanding of cortex? Or do they show two different aspects of the same great brain region?

Regardless of one's assumptions, multiple lines of evidence show that the cortex (together with thalamus) is able to identify, record and act on conscious perceptual information with extraordinary accuracy, down to the physical minima of receptor activation by light, air vibrations, chemical and skin receptors. More than two centuries of psychophysics demonstrates the precision of conscious perception.

11.1.2 Cortical Intention Processing

Yet cortex also supports *intention processing* in pursuit of the Darwinian imperatives of species survival and reproduction.

These two aspects are arguably complementary. Accurate perception is pragmatically essential—as in the case of olfactory spoor-following with *friend or foe* identification, in species like rodents, dogs and cats. Epistemic accuracy may not be the opposite of pragmatic readiness. In many cases accurate perception and pursuit of intentions may be symbiotic.

Freeman and Kozma have shown that pragmatic epistemology can be remarkably useful in revealing mass action phenomena in cortex, by focusing on the dendritic meshwork of the surface layer of neocortex and paleocortex. They show that after Hilbert analysis, a strip of cortical electrodes with sufficient spatial density reveals classical nonlinear dynamical phenomena.

This essay suggests that Freeman Neurodynamics may be complementary to conventional analysis of ECoG. We suggest that the highly distributed multiscale activity revealed by Freeman Neurodynamics supports accurate input identification by way of sparse distributed memory [7, 13]. This entails a prediction, namely that some brain region acts as a radically condensed SDM vector for synaptic trace deposition and retrieval in cortex, for episodic (conscious) learning. Possible SDM vector substrates may exist in the hippocampus, claustrum, colliculi or even small thalamic nuclei.

11.1.3 Freeman Neurodynamics

Freeman Neurodynamics and conventional ECoG analysis use different analytical methods, showing very different patterns of activity emanating from the same neurons. After all, the horizontal dendritic feltwork of Layer I reflects the same pyramidal neurons that send and receive axons on vertical trajectories.

The cerebral cortex, *in resonance with* thalamic nuclei, supports conscious perception, in marked contrast to other brain structures like the cerebellum. Multiple lines of evidence link cortex to specifically conscious sensory events, including direct cortical stimulation in waking subjects, precise lesions in sensory regions, and cortical task recordings in waking primates. Studies in the macaque, with a similar visual system to ours, converge well with a large body of human brain experiments.[1]

---

[1] Some regions of cortex do not support conscious contents directly, like the parietal maps of nearby space, as shown by visual neglect. Therefore cortex seems to be needed for conscious perception, but there are cortical regions that do not (directly) support conscious perception. I have made the argument that parietal egocentric and allocentric visual maps provide contextual information to define the nearby space within which conscious visual objects and events are experienced. Thus parietal regions (dorsal visual stream) are needed for visual percepts, but do not directly support the visible features of percepts.

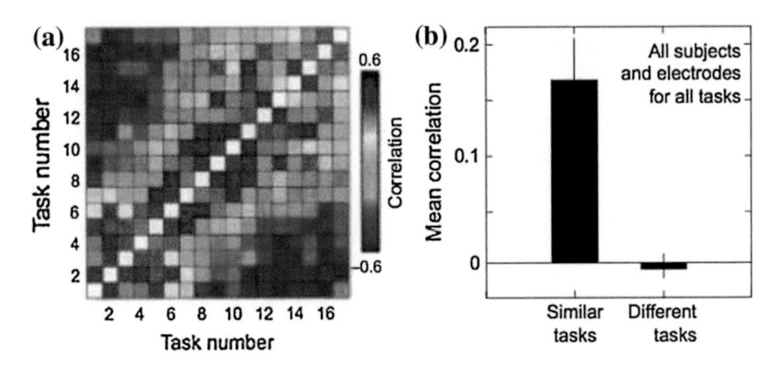

Fig. 11.1 Theta-gamma phase-locking among 16 different tasks, from [3]

11.2 Binocular Rivalry in Primates

There are now many studies using methods like binocular rivalry to deliver two similar (but not identical) stimuli to each eye, such that the two inputs cannot be fused into a single gestalt. Only one of the two input streams can be reported *as conscious* at any time, allowing for cortical recording of the two processing streams, conscious and unconscious. The Logothetis laboratory has used this approach to study conscious vision in the macaque for more than two decades.

The standard textbook account describes visual cortex as more than 40 visuotopical and spatiotopical arrays, linked by bi-directional axons that preserve *labeled line* connectivity. Active excitatory cells in each successive visual feature array excite corresponding cells in other arrays. Binocular rivalry leads to lateral competition in one or more levels of the visual hierarchy, while binocular fusion reflects mutually supportive activity. A winner-take-all coalition of active neurons in multiple visuotopical arrays leads to convergent resonance (binding) of active cells of the hierarchy, followed by widespread gestalt propagation [2].

Since spatiotopical organization is common in cortex, visual gestalt *broadcasting* may excite spatial maps in somatotopical, audiospatial and somatomotor regions. In principle, this could also explain why conscious stimuli are voluntarily reportable, using visuotopical maps in executive and motor regions of the frontal lobe [2].

Figure 11.1 shows how intracranial recording in waking humans reveals task-specific theta-gamma phase-locking in cortex [3]. Phase-locking between *similar* tasks is shown near the diagonal, with an average $r \sim 0.17$, while the correlation between *different* tasks show $r \sim 0$. The difference is statistically significant.

11.3 Dynamic Global Workspace Theory

A recent paper by Kozma and Freeman suggests significant points of contact between Freeman Neurodynamics and Dynamic Global Workspace Theory (d-GWT) [2, 10]. They write: "Recent brain monitoring experiments indicate intermittent singularities in cognitive processing. We employ the biologically motivated Freeman K models to interpret these findings. In particular, we show that random graph theory provides a suitable mathematical framework to describe the experimentally-observed singularities as phase transitions across the cortical neuropil. We introduce the hypothesis that the rapid propagation of phase dispersion over the hemisphere is the manifestation of the cognitive broadcast as described in Baars' global workspace theory. In addition, our exponentially expanding brain graph model using pioneer neuron sub-plates can be used for describing recent findings on the presence of rich club structures in brain networks."

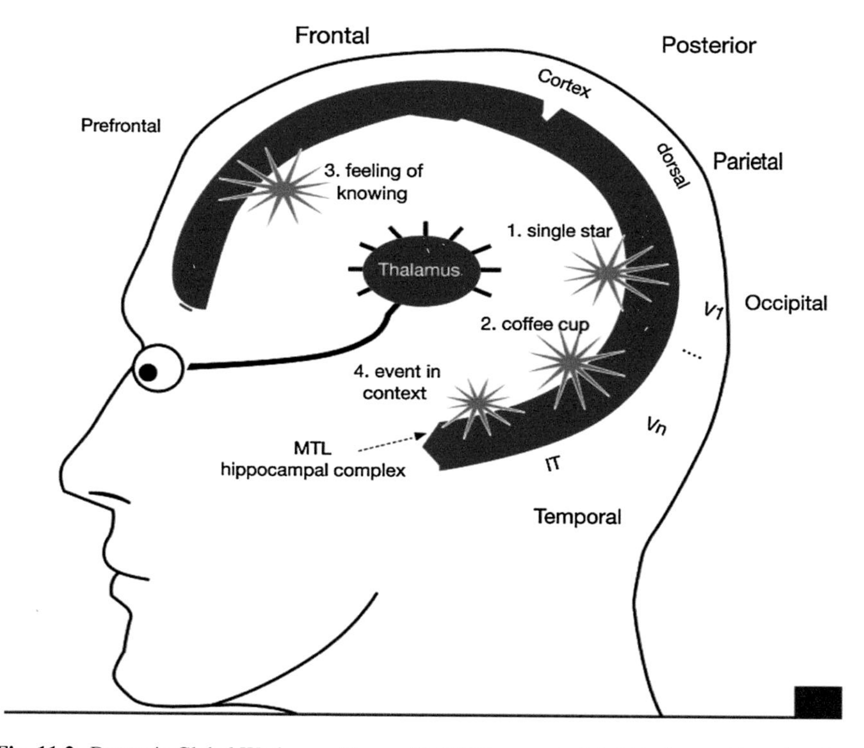

Fig. 11.2 Dynamic Global Workspace Theory (d-GWT) suggest global binding and broadcasting involving gestalt convergence in any region of cortex, such as visual areas V1, IT, and MTL propagating to any other region(s). Different levels of perception are proposed to bind and propagate from different visual regions. C-T axonal signaling is bidirectional between cortical columns, suggesting that *reentrant* or *adaptive resonance* is the norm. Cortex reveals a wide range of oscillatory phenomena, rather than unidirectional signal transmission, as in electronics

In d-GWT, sensory input can evoke a resonant cell assembly that can bind into a coherent gestalt after winner-take-all competition involving any set of activated cortical arrays; likewise, global broadcasting propagates from the gestalt formation to any other cortical regions. Such any-to-any signaling in both gestalt binding and broadcasting is claimed to account for the great flexibility of the cortico-thalamic (C-T) complex in mammalian brains. In Dynamic Global Workspace Theory *binding and broadcasting* can occur in any part of cortex, Fig. 11.2.

11.3.1 Direct Evidence for Cortical Binding and Broadcasting

A recent report from the Logothetis group gives the most direct evidence to date for *global binding and broadcasting* in the conscious (reportable) visual stream. Panagiotaropoulos et al. [12] show that in the macaque, the reportable stream settles on a coherent gestalt high in the visual hierarchy (MTL); gestalt features then propagate forward to prefrontal visual arrays.

In contrast, unconscious visual input does not *bind and broadcast* in this way.

The Panagiotaropoulos et al. experiment was able for the first time to demonstrate accurate broadcasting [12], since the same visual features were active in both visual

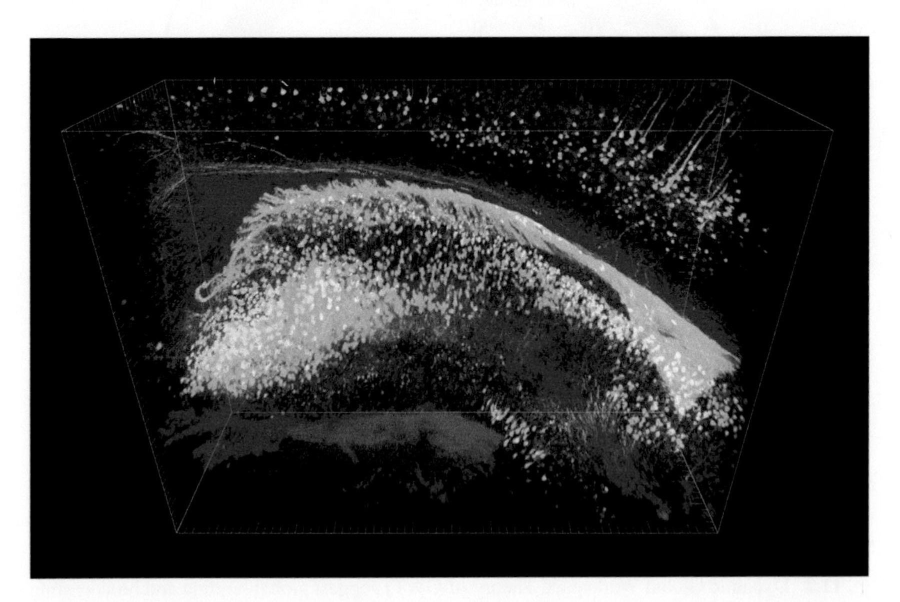

Fig. 11.3 Chung and Deisseroth [4] show a remarkably clear photo of preserved mouse hippocampus using the CLARITY method for substituting a clear aerogel for myelinated glial cells that normally obscure neurons. Cellular features are color enhanced. Pyramidal cell bodies are clearly visible in different hippocampal layers (*colored dots*). The dendritic feltwork of the surface layer of hippocampus (CA1) appears as a *thin white* horizontal meshwork. Hippocampus is three-layered paleocortex, but its major features also appear in six-layered neocortex

cortex and PFC (executive and goal integration), as predicted by global workspace theory in 1988 [1].

Cortical recording shows a wide range of oscillatory phenomena, including turbulent wavefronts, vortices, standing and traveling waves, cross-frequency phase linking, cross-regional phase linking and delinking, and oscillatory grouping, ranging from Slow Oscillations (<0.1 Hz) to high gamma (200 Hz). Theta-gamma, alpha-beta, and alpha-gamma linking seem to be common.

Because topographical arrays are common in cortex, these wave phenomena are thought to reflect spatiotemporal signaling between different arrays of cortical neurons (Fig. 11.3).

11.4 Freeman Neurodynamics

Freeman [6] used Hilbert analysis to process the electrical field output of a set of cortical electrodes, revealing classical nonlinear phenomena, including ~100 ms phase equilibria followed by fast chaotic collapses (~5 ms), see Fig. 11.4. These cycling equilibria and chaotic collapses near 10 Hz may spread to large cortical territories, ranging from a few square centimeters to an entire hemisphere.

Hilbert analysis is plausible for Layer I of the cortex, viewed as a spatial input analyzer, consisting of a horizontal meshwork of apical dendrites that appears histologically quite different from the (vertical) pyramidal resonance discussed above. Yet the horizontal dendritic feltwork must also interact y with vertical pyramidal firing,

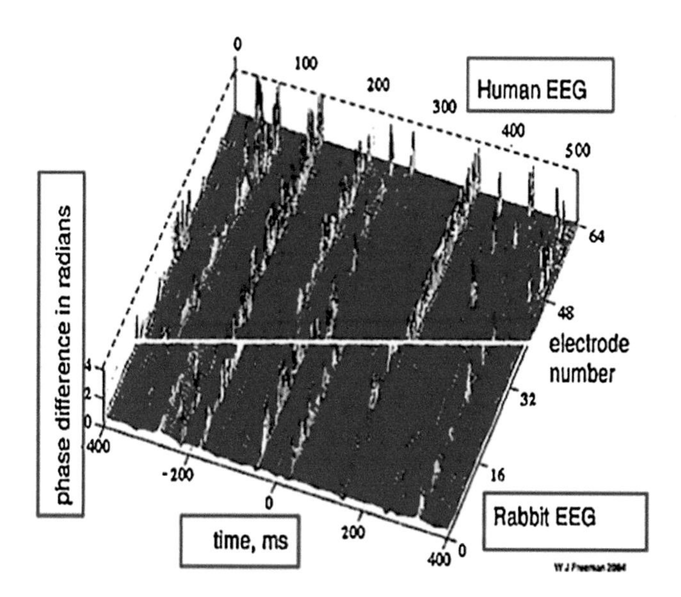

Fig. 11.4 Freeman Neurodynamics: Nonlinear chaotic phenomena in cortex; from [6]

since dendrites and axons belong to the same neurons. Dendritic voltages alter the probability of axonal firing. Axonal spikes momentarily change dendritic membrane potentials.

Freeman and coworkers' results are qualitatively different from textbook descriptions of cortical signaling [6, 8, 9]. Equilibria and chaotic collapses typify a set of deterministic but nonlinear dynamical systems, such as phase changes in the organization of water, sand piles, and weather forecasting [11].

11.5 An Integrative Hypothesis

We suggest a tentative integrative hypothesis, namely that the highly distributed multi-scale activity revealed by Freeman Neurodynamics supports accurate input identification by way of sparse distributed memory [7, 13]. This entails a prediction that some brain region acts as a radically condensed SDM vector for memory deposition and retrieval. The hippocampus or claustrum may be candidates for an SDM vector substrate.

11.5.1 Reference Notes

This paper reflects many discussions with Walter J. Freeman and Stan Franklin, to whom it is dedicated. Stan helped to clarify the distinctive properties of Kanerva's Sparse Distributed Memory (SDM), which have been further developed by Franklin and Javier Snaider (see references). I am also grateful to Robert Kozma for an advance copy of Kozma and Freeman [10], which helped me to clarify and focus on points of contact between Freeman Neurodynamics and global workspace theory. Needless to say, any remaining errors are my own.

References

1. Baars BJ (1988) A cognitive theory of consciousness. Cambridge University Press, New York
2. Baars BJ, Franklin S, Ramsoy TZ (2013) Global workspace dynamics: cortical binding and propagation enables conscious contents. Front Psychol 4:200. doi:10.3389/fpsyg.2013.00200. eCollection
3. Canolty RT, Edwards E, Dalal SS, Soltani M, Nagarajan SS, Kirsch HE, Berger MS, Barbaro NM, Knight RT (2006) High gamma power is phase-locked to theta oscillations in human neocortex. Science 313(5793):1626–1628
4. Chung K, Deisseroth K (2013) CLARITY for mapping the nervous system. Nat Methods 10(6):508–513. doi:10.1038/nmeth.2481
5. Franklin SP (1997) Artificial minds. Bradford Books/MIT Press, Cambridge
6. Freeman WJ (2003) Evidence from human scalp electroencephalograms of global chaotic itinerancy. Chaos 13(3):1067–1077

7. Kanerva P (2003) Sparse distributed memory. Bradford Books/MIT Press, Cambridge
8. Kozma R, Puljic M, Balister P, Bollobas B, Freeman WJ (2005) Phase transitions in the neuropercolation model of neural populations with mixed local and non-local interactions. Biol Cybern 92(6):367–379
9. Kozma R (2007) Neuropercolation. Scholarpedia 2(8):1360
10. Kozma R, Freeman WJ (2015) Modeling cortical singularities during the cognitive cycle using random graph theory. Advances in cognitive neurodynamics (IV). Springer, The Netherlands, pp 137–142
11. Lorenz EN (1969) Three approaches to atmospheric predictability. Bull. Am. Meteorol. Soc. 50:345–349
12. Panagiotaropoulos TI, Deco G, Kapoor V, Logothetis NK (2012) Neuronal discharges and gamma oscillations explicitly reflect visual consciousness in the lateral prefrontal cortex. Neuron 74(5):924–935. doi:10.1016/j.neuron.2012.04.013
13. Snaider J, Franklin S (2014) Modular composite representation. Cogn. Comput. 6(3):510–527. doi:10.1007/s12559-013-9243-y

Chapter 12
Commentary by Steven L. Bressler

Interareal Neocortical Actions By Neuronal Populations

Steven L. Bressler

12.1 Introduction

The purpose of this short chapter is to advance our understanding of the functional actions that occur between different areas of the mammalian neocortex. This topic is of immense importance to the question of the neural basis of cognition, both in animals and humans. For example, one conceptual framework that has emerged over many years describes cognitive function in terms of actions between areas in large-scale networks of the neocortex, or neurocognitive networks [12, 36–38]. It posits the neocortical area as a computational processing entity in the brain, and the large-scale structure of anatomical pathways connecting those areas as the connectivity matrix that determines the interareal actions underlying the cortical computations of a species. The inhomogeneity of the large-scale cortical connectivity suggests that the neocortex is not a homogeneous computational medium and that interareal cortical connectivity is important for cortical function. The neurocognitive network framework is directly linked to the idea that cognition is a collective phenomenon that emerges from the actions exerted between cortical areas. In short, the nature of cognition in the human brain is intimately tied to an understanding of how cortical areas act upon one another, and how those actions lead to emergent neurocognitive phenomena.

The actions brought to bear between neocortical areas are no doubt synaptic and excitatory, and the excitatory synapses are found at the terminals of fiber pathways that project from a sending area to a receiving area. These interareal pathways consist of axons from the large projection neurons in the sending areas. These are the excitatory pyramidal cells, which are the cellular generators of many of the observable macroscopic neural fields of the brain, such as the electroencephalogram (EEG), magnetoencephalogram (MEG), electrocorticogram (ECoG), and local field potential (LFP) [35]. The cell types in the areas that receive the pyramidal cell axonal terminals are multiple and variable [6, 7, 43]. Interareal feedforward excitation occurs when

S. Bressler
Center for Complex Systems and Brain Sciences, Florida Atlantic University, Boca Raton, FL, USA
e-mail: bressler@fau.edu; bressler.stevenl@gmail.com

R. Kozma and W.J. Freeman, *Cognitive Phase Transitions in the Cerebral Cortex – Enhancing the Neuron Doctrine by Modeling Neural Fields*, Studies in Systems, Decision and Control 39, DOI 10.1007/978-3-319-24406-8_12

projection neurons excite recipient neurons (either other projection neurons or excitatory interneurons), and interareal feedforward inhibition occurs when the projection neurons excite recipient inhibitory interneurons, thus feeding inhibition into the receiving area. For projections between hierarchically arranged areas such as in visual cortex, the termination layer pattern depends on whether the projection is bottom-up, top-down, or lateral [24]. Since the pathways between cortical areas are predominantly bi-directional, the actions are usually exerted concurrently in both directions. Hence, each area is typically both sending and receiving, and bi-directional cortical interactions are much more prevalent than uni-directional actions.

In this chapter, I will contrast evidence for two radically different models of the actions exerted by neurons in a sending cortical area on those in a receiving area. The first (neuron–neuron) model considers that actions (or interactions) occur between individual cortical neurons, whereas the second considers that they occur between neuronal populations. From the well established neuron doctrine [8, 50], it often seems intuitive to suppose that single neurons are the basis for interareal actions in the neocortex, and hence that the neuron–neuron model is the correct one. However, although the individual cortical neuron has repeatedly been shown to signal specific sensory features, it does not necessarily follow that the transmission of sensory information from one cortical area to another transpires, point-to-point, from one cortical neuron to another. Hence, the second (population–population) model, although counterintuitive, may be the correct one.

The distinction between the neuron–neuron and the population–population models may appear at first glance to rest on the difference between single-neuron and population coding of perceptual and cognitive features. However, interaction models are independent of coding models, and it may be that single neurons perform coding within cortical areas whereas the transmission of cognition-relevant activity between areas is by neuronal populations. Similarly, local, within-area processing may be performed by the neuronal assembly, whereas the action exerted between areas may be by way of single neurons. In any case, the question under consideration is not whether neurons carry influences between neocortical areas. Both the neuron–neuron and the population–population models hold that they do. Rather, the question is whether interareal interactions are organized at the level of the single neuron or the neuronal population.

The prevalence of interareal functional effects observed in the cortex [30, 31, 56] suggests that interareal interactions are important for cortical computation. However, it is not known whether the interactions are organized at the single neuron or neuronal population levels. One way to approach this question is by comparison of the evidence for interareal neuron–neuron and population–population interactions. If there is ample evidence for interareal neuron–neuron but not population–population interactions, then the neuron–neuron model is to be preferred. If there is ample evidence for interareal population–population but not neuron–neuron interactions, or there is ample evidence for both, then the population–population model is favored.

In addition to the spatial scale of interaction (neuron–neuron or population–population), the temporal scale is also relevant. The temporal scale refers to the length of time over which the interaction is in effect. To be relevant for neocortical function, significant interactions must transpire on a time scale commensurate with that of functional events. Since these events may be brief (i.e. on a sub-second time scale), evidence must exist that interareal interactions are also brief.

12.2 Neuron–Neuron Interactions

Neuron–neuron interactions in the neocortex are typically studied by cross-correlation analysis of simultaneously recorded multiple neuron spike trains [2]. Similar analyses focus on detecting repeated firing patterns in multiple-neuron spike-train data that repeat with above-chance likelihood [1]. Cross-correlation analysis reveals neuron–neuron spike synchrony that is brief and significantly above the background level. Often, neuron–neuron synchronous firing is observed to occur briefly in sensory cortex in response to a sensory stimulus or briefly in motor cortex in relation to a movement: Both sensory and motor event-related neuron–neuron synchronies are well established and have important functional relations [2, 3, 44]. For example, the task-related synchronous firing of auditory cortical neurons reflects the strengthening of functional synaptic connections between them [4], and neuron pairs in prefrontal cortex show briefly correlated activity in relation to working memory processes [27].

There have been numerous reports of stimulus-evoked briefly synchronous firing between neurons in different visual cortical areas of the cat [20–22, 39] and monkey [15, 40, 46]. Inter-neuron coupling is most often observed between cells having overlapping receptive fields during co-activation by visual stimuli [53, 54]. Although neurons with the same orientation selectivity in different visual areas are not preferentially linked [39], the incidence of synchronous firing is greater for cells having overlapping receptive fields and similar orientation preference [53].

Synchronous interareal firing is usually weak except when neuronal activity is oscillatory, implying that interareal synchrony depends on population activity. This is because oscillations are thought to arise in neuronal activity due to the interactions of intra-areal neuronal populations, not individual neurons. Thus, although individual cortical neurons display oscillatory activity, it is because they are imbedded in intra-areal neuronal populations that oscillate; the synaptic connectivity that is responsible for the oscillation is a population property that is shared by the neurons in the population [25]. This explanation can account for the finding that stimulus-induced coupling of V1–V2 spiking activity depends on the presence of a strong and coherent gamma rhythm in these visual cortical areas [32]. The implication for interareal interaction is that interareal neuron–neuron interaction depends on intra-areal population activity, supporting the idea that interareal interactions are organized at the population level.

12.3 Population–Population Interactions

That interareal cortical interactions are organized at the level of neuronal populations is suggested by the high degree of divergence and convergence, and the high level of density, in the anatomical connectivity between cortical areas. Because the axons of single projection neurons branch profusely in receiving areas, each projection neuron provides excitatory drive to multiple recipient neurons and each recipient neuron receives excitatory drive from multiple projection neurons. Furthermore, because of the high density of connectivity, the divergence and convergence ratios are extremely high. Therefore, based solely on anatomical projection patterns, interareal cortical interactions are most likely to be organized at the level of neuronal populations.

Interactions between neuronal populations in different areas of the neocortex are typically measured by the cross-correlation or spectral coherence between LFPs simultaneously recorded from the areas. The LFP arises in the neocortex as dendritic potentials from neuronal populations [35]. It is generally agreed that the LFP is not transmitted between areas (although see [5, 33, 51]). However, the LFP at a cortical recording site may serve as an index of the joint neuronal activity of the population located at that site that is more readily recorded with cortical microelectrodes than unit activity. Also, since the LFP is more highly correlated with the fMRI BOLD signal than unit activity [34], the LFP more closely reflects the neurovascular response of a neuronal population in the neocortex.

The cross-correlation or spectral coherence between LFPs simultaneously recorded from two cortical areas thus represents a measure of the interaction between those areas, even though the population spiking activity that carries the influences between areas is not directly recorded. Since the LFP is thought to reflect the input to, and intracortical processing of, an area, and spiking activity the output, the spike-field coherence, with simultaneous recording of spiking in one area and the field (LFP) in the other, is also used to measure interareal cortical interactions [41, 48].

One of the first reports on the functional relevance of interareal cortical interaction was in the paleocortex, where oscillatory synchrony between simultaneously recorded LFPs from the olfactory bulb and cortex was found in relation to respiratory inspiration [10]. Significant spatial variation across both the bulb and cortex was seen in the degree of bulbo-cortical interaction. Feedforward driving of the cortex by the bulb was only partly responsible for bulbo-cortical interaction during inspiration [11]. Other factors considered were convergence of transmission from distributed bulbar populations to a single cortical population, independent self-generation of the cortical LFP, cortex-to-bulb feedback, and common-input driving of both bulb and cortex by an outside source. All of these factors are also expected to play a role in population-based interareal neocortical interactions.

Subsequently, a number of studies have used the cross-correlation or spectral coherence of simultaneously recorded LFPs to demonstrate the functional relevance of interareal neocortical interactions: interareal synchrony between visual and motor cortical areas indexes visuomotor behavior [13]; interareal visual cortical synchrony reflects stimulus expectancy [55]; interareal cortical synchrony in the dorsal visual pathway reflects spatial attention [47]; interareal synchrony between pre- and post-central cortical areas reveals a distributed large-scale sensorimotor network related to the maintenance of steady arm and hand muscle contractions [14]; the magnitude and frequency of synchrony between prefrontal and posterior parietal cortical areas distinguish bottom-up from top-down forms of attention [16]; the synchrony between prefrontal and posterior parietal cortical areas distinguishes the content of visual working memory [48]; the synchrony between visual areas V1 and V2 reflects contrast stimulation [45]; and interareal visual synchrony reflects attentional stimulus selection [9]. Likewise, spike-field coherence demonstrates interareal interaction between posterior parietal and premotor cortices in relation to free choice [41] and between posterior parietal and prefrontal cortices in relation to visual working memory [48].

Another line of evidence that relates to interareal interaction comes from studies of cryogenic blockade. Generally, the reversible cooling of one cortical area affects the function of neurons in another area. Cells in visual cortical area 17 of the squirrel monkey become less responsive to visual stimulation when area 19 is cooled [49]; cooling of either the dorsolateral prefrontal cortex or the inferotemporal cortex lowers spontaneous neuronal firing, and that related to a visual short-term memory task, in the other area [28], cooling of posterior parietal cortex lowers firing rates of prefrontal neurons [42]; cooling of either posterior parietal or dorsolateral prefrontal cortex impairs neuronal firing related to an oculomotor delayed response task in the other area [18]; and cooling of either primary or secondary auditory cortex suppresses neuronal responses to tonal stimulation in the other area [17]. Since the cooling in these studies is applied to an entire cortical area, the results appear to support the population–population model of interaction between cortical areas. However, the neuron–neuron interaction model cannot be ruled out because cooling of an entire area affects the individual neurons within it.

12.4 Discussion

The question of how neocortical areas act upon one another is of immense importance for understanding the neural basis of cognition. It is well established that the neocortex is critically important for cognitive functions. To realize the veracity of this point, one need only consider the devastating cognitive impairments that result from cortical lesions. Many brain theories of cognition rely on coherent interactions between different cortical areas [19, 23, 26, 29, 52].

The bulk of currently available evidence, both from neuroanatomy and neurophysiology, suggests that the interactions between cortical areas are organized at the level of the neuronal population. Studies of correlated spiking from neurons in different areas of the neocortex show that synchronous interareal firing is only robust when neuronal population activity is oscillatory. This finding suggests that interareal neuron–neuron cortical interactions are embedded in oscillatory population–population interactions, and that interareal neuron–neuron correlations result from population–population correlations. In contrast to the evidence for interareal neuron–neuron interactions, the evidence for interareal population–population interactions, suggested by ample studies of interareal field-field and spike-field correlations, and by interareal cortical cooling effects, is robust.

References

1. Abeles M, Bergman H, Margalit E, Vaadia E (1993) Spatiotemporal firing patterns in the frontal cortex of behaving monkeys. J Neurophysiol 70:1629–1638
2. Aertsen A, Arndt M (1993) Response synchronization in the visual cortex. Curr Opin Neurobiol 3:586–594
3. Aertsen A, Gerstein GL (1985) Evaluation of neuronal connectivity: sensitivity of cross-correlation. Brain Res 340:341–354
4. Ahissar E, Vaadia E, Ahissar M, Bergman H, Arieli A, Abeles M (1992) Dependence of cortical plasticity on correlated activity of single neurons and on behavioral context. Science 257:1412–1414
5. Alle H, Geiger JRP (2006) Combined analog and action potential coding in hippocampal mossy fibers. Science 311:1290–1293
6. Barbas H (1993) Organization of cortical afferent input to orbitofrontal areas in the rhesus monkey. Neuroscience 56:841–864
7. Barbas H (2000) Connections underlying the synthesis of cognition, memory, and emotion in primate prefrontal cortices. Brain Res Bull 52:319–330
8. Barlow HB (1972) Single units and sensation: a neuron doctrine for perceptual psychology. Perception 1:371–394
9. Bosman CA, Schoffelen J-M, Brunet N, Oostenveld R, Bastos AM, Womelsdorf T, Rubehn B, Stieglitz T, De Weerd P, Fries P (2012) Attentional stimulus selection through selective synchronization between monkey visual areas. Neuron 75:875–888
10. Bressler SL (1987a) Relation of olfactory bulb and cortex. I. Spatial variation of bulbocortical interdependence. Brain Res 409:285–293
11. Bressler SL (1987b) Relation of olfactory bulb and cortex. II. Model for driving of cortex by bulb. Brain Res 409:294–301
12. Bressler SL (2008) Neurocognitive netw Schol 3:1567
13. Bressler SL, Coppola R, Nakamura R (1993) Episodic multiregional cortical coherence at multiple frequencies during visual task performance. Nature 366:153–156
14. Brovelli A, Ding M, Ledberg A, Chen Y, Nakamura R, Bressler SL (2004) Beta oscillations in a large-scale sensorimotor cortical network: directional influences revealed by Granger causality. Proc Natl Acad Sci USA 101:9849–9854
15. Bullier J, Munk MHJ, Nowak LG (1992) Synchronization of neuronal firing in areas V1 and V2 of the monkey. Soc Neurosci Abstr 18(11):7
16. Buschman TJ, Miller EK (2007) Top-down versus bottom-up control of attention in the prefrontal and posterior parietal cortices. Science 315:1860–1862
17. Carrasco A, Lomber SG (2010) Reciprocal modulatory influences between tonotopic and nontonotopic cortical fields in the cat. J Neurosci 30:1476–1487
18. Chafee MV, Goldman-Rakic PS (2000) Inactivation of parietal and prefrontal cortex reveals interdependence of neural activity during memory-guided saccades. J Neurophysiol 83:1550–1566
19. Dehaene S, Naccache L (2001) Towards a cognitive neuroscience of consciousness: basic evidence and a workspace framework. Cognition 79:1–37
20. Eckhorn R, Bauer R, Jordan W, Brosch M, Kruse W et al (1988) Coherent oscillations: a mechanism for feature linking in the visual cortex. Biol Cybern 60:121–130

21. Engel AK, Kreiter AK, Konig P, Singer W (1991) Synchronization of oscillatory neuronal responses between striate and extrastriate visual cortical areas of the cat. Proc Natl Acad Sci USA 88:6048–6052
22. Engel AK, Konig P, Kreiter AK, Singer W (1991) Interhemispheric synchronization of oscillatory neuronal responses in cat visual cortex. Science 252:1177–1179
23. Fan J, Posner M (2004) Human attentional networks. Psychiatr Prax 31:S210–S214
24. Felleman DJ, Van Essen DC (1991) Distributed hierarchical processing in the primate cerebral cortex. Cereb Cortex 1:1–47
25. Freeman WJ (2005) A field-theoretic approach to understanding scale-free neocortical dynamics. Biol Cybern 92:350–359
26. Fries P (2005) A mechanism for cognitive dynamics: neuronal communication through neuronal coherence. Trends Cogn Sci 9:474–480
27. Funahashi S, Inoue M (2000) Neuronal interactions related to working memory processes in the primate prefrontal cortex revealed by cross-correlation analysis. Cereb Cortex 10:535–551
28. Fuster JM, Bauer RH, Jervey JP (1985) Functional interactions between inferotemporal and prefontal cortex in a cognitive task. Brain Res 330:299–307
29. Horwitz B, Braun AR (2004) Brain network interactions in auditory, visual and linguistic processing. Brain Lang 89:377–384
30. Hupé JM, James AC, Payne BR, Lomber SG, Girard P, Bullier J (1998) Cortical feedback improves discrimination between figure and background by V1, V2 and V3 neurons. Nature 394:784–787
31. Hupé JM, James AC, Girard P, Lomber SG, Payne BR, Bullier J (2001) Feedback connections act on the early part of the responses in monkey visual cortex. J Neurophysiol 85:134–145
32. Jia X, Tanabe S, Kohn A (2013) Gamma and the coordination of spiking activity in early visual cortex. Neuron 77:762–774
33. Kruger L, Otis TS (2007) With withered Golgi? A retrospective evaluation of reticularist and synaptic constructs. Brain Res Bull 72:201–207
34. Logothetis NK, Pauls J, Augath M, Trinath T, Oeltermann A (2001) Neurophysiological investigation of the basis of the fMRI signal. Nature 412:150–157
35. Lopes da Silva F (2013) EEG and MEG: relevance to neuroscience. Neuron 80:1112–1128
36. Meehan T, Bressler SL (2012) Neurocognitive networks: findings, models, and theory. Neurosci Biobehav Rev 36:2232–2247
37. Mesulam MM (1990) Large-scale neurocognitive networks and distributed processing for attention, language, and memory. Ann Neurol 28:597–613
38. Mesulam MM (1998) From sensation to cognition. Brain 121:1013–1052
39. Nelson JI, Salin PA, Munk MHJ, Arzi M, Bullier J (1992) Spatial and temporal coherence in cortico-cortical connections: a cross-correlation study in areas 17 and 18 in the cat. Vis Neurosci 9:21–38
40. Nowak LG, Munk MHJ, Chounlamountri N, Bullier J (1994) Temporal aspects of information processing in areas V1 and V2 of the macaque monkey. In: Pantev C (ed) Oscil Event-Related Brain Dyn. Plenum Press, New York, pp 85–98
41. Pesaran B, Nelson MJ, Andersen RA (2008) Free choice activates a decision circuit between frontal and parietal cortex. Nature 453:406–409
42. Quintana J, Fuster JM, Yajeya J (1989) Effects of cooling parietal cortex on prefrontal units in delay tasks. Brain Res 503:100–110
43. Rempel-Clower NL, Barbas H (2000) The laminar pattern of connections between prefrontal and anterior temporal cortices in the Rhesus monkey is related to cortical structure and function. Cereb Cortex 10:851–865
44. Riehle A, Grun S, Diesmann M, Aertsen A (1997) Spike synchronization and rate modulation differentially involved in motor cortical function. Science 278:1950–1953
45. Roberts MJ, Lowet E, Brunet NM, Ter Wal M, Tiesinga P, Fries P, De Weerd P (2013) Robust gamma coherence between macaque V1 and V2 by dynamic frequency matching. Neuron 78:523–536

46. Roe AW, Tso DY (1992) Functional connectivity between V1 and V2 in the primate. Soc Neurosci Abstr 18(11):4
47. Saalmann YB, Pigarev IN, Vidyasagar TR (2007) Neural mechanisms of visual attention: how top-down feedback highlights relevant locations. Science 316(5831):1612–1615
48. Salazar RF, Dotson NM, Bressler SL, Gray CM (2012) Content-specific fronto-parietal synchronization during visual working memory. Science 338:1097–1100
49. Sandell JH, Schiller PH (1982) Effect of cooling area 19 on striate cortex cells in the squirrel monkey. J Neurophysiol 48:38–48
50. Shepherd GM (1991) Found neuron doctrine. Oxford University Press, New York
51. Shu Y, Hasenstaub A, Duque A, Yu Y, McCormick DA (2006) Modulation of intracortical synaptic potentials by presynaptic somatic membrane potential. Nature 441:761–765
52. Singer W (1994) Coherence as an organizing principle of cortical functions. Int Rev Neurobiol 37:153–183
53. Singer W, Gray CM (1995) Visual feature integration and the temporal correlation hypothesis. Annu Rev Neurosci 18:555–586
54. Smith MA, Jia X, Zandvakili Kohn A (2013) Laminar dependence of neuronal correlations in visual cortex. J Neurophysiol 109:940–947
55. von Stein A, Chiang C, Konig P (2000) Top-down processing mediated by interareal synchronization. Proc Natl Acad Sci USA 97:14753–14758
56. Young MP (2000) The architecture of visual cortex and inferential processes in vision. Spat Vis 13:137–146

Chapter 13
Commentary by Zoltán Somogyvári and Péter Érdi

Forward and Backward Modeling: From Single Cells to Neural Population and Back

Zoltán Somogyvári and Péter Érdi

Abstract Some aspects of forward and backward neural modeling are discussed, showing, that the neural mass models may provide a "golden midway" between the detailed conductance based neuron models and the oversimplified models, dealing with the input–output transformations only. Our analysis combines historical perspectives and recent developments concerning neural mass models as a third option for modeling large neural populations and inclusion of detailed anatomical data into them. The current source density analysis and the geometrical assumption behind the different methods, as an inverse modeling tool for determination of the sources of the local field potential is discussed, with special attention to the recent results about source localization on single neurons. These new applications may pave the way to the emergence of a new field of micro-electric imaging.

13.1 Modeling Population of Neurons: The Third Option

Structure-based bottom-up modeling has two extreme alternatives, namely **multi-compartmental** simulations, and simulation of **networks** composed of simple elements. There is an obvious trade-off between these two modeling strategies. The first method is appropriate to describe the electrogenesis and spatiotemporal

Z. Somogyvári · P. Érdi
Department of Theory, Wigner Research Center for Physics of the Hungarian Academy of Science, Konkoly-Thege Miklós út 29-33, 1121 Budapest, Hungary

Z. Somogyvári
Department of Epilepsy and General Neurology, National Institute of Clinical Neurosciences, Budapest, Hungary

P. Érdi
Center for Complex System Studies, Kalamazoo College, 1200 Academy Street, Kalamazoo, Mi 49006, USA

R. Kozma and W.J. Freeman, *Cognitive Phase Transitions in the Cerebral Cortex – Enhancing the Neuron Doctrine by Modeling Neural Fields*, Studies in Systems, Decision and Control 39, DOI 10.1007/978-3-319-24406-8_13

propagation of the action potential in single cells, and in small and moderately large networks based on data on detailed morphology and kinetics of voltage- and calcium-dependent ion channels. The mathematical framework is the celebrated Hodgkin–Huxley model [19] supplemented with the cable theory [34, 35]. The construction of neural simulation softwares such as NEURON [17, 18], and GENESIS [5] contributed very much to make the emerging field of computational neuroscience is able to make realistic bottom up neural simulations. The second approach grew up from the combination of the McCulloch–Pitt neuron models and the of the Hebbian learning rule, and offers a computationally efficient method for simulating large network of neurons where the details of single cell properties are neglected. A classical example of using two-level neural network models by combining activity and synaptic dynamics as a model of generating ordered neural pattern by a self-organizing algorithm is [47], and a newer one for invariant pattern recognitions in the same spirit [4].

As concerns single cell modeling, there is a series of cell models with different level of abstraction. While multi-compartmental models take into account the spatial structure of a neuron, neural network techniques are generally based on integrate-and-fire models. The latter is a spatially homogeneous, spike-generating device. For a review of 'spiking neurons' see [13]. As is well known, neural network theory, incorporating biologically non-plausible learning rules became a celebrated subclass of machine learning discipline called *artificial neural network*. Modeling population of neurons emerged as a compromise between "too microscopic" and "too macroscopic" descriptions [10].

13.2 Mesoscopic Neurodynamics

13.2.1 Statistical Neurodynamics: Historical Remarks

There is a long tradition to try to connect the 'microscopic' single cell behavior to the global 'macrostate' of the nervous system analogously to the procedures applied in statistical physics. Global brain dynamics is handled by using continuous (neural field) description instead of the networks of discrete nerve cells. Both deterministic, field-theoretic [1, 15, 37, 46] and more statistical approaches have been developed. [10] introduced a modular and therefore hierarchical framework of neural field models, as the series of K models from K0 to KIII. This series of more and more complex neural mass models has been reached the level of behavioral analysis with the KIII sets. Later, [25] extended the K sets theory to the next (KIV) level to account for the interaction between cortical areas as well. One of us (PE) participated in the application of KIV system for hippocampus-related problems [26, 27].

This way, the otherwise pure statistical handling of neural populations gains new anatomical details.

Francesco Ventriglia constructed a neural **kinetic theory** of large-scale brain activities that he presented in a series of somewhat overlooked papers [41–44]. His statistical theory is based on two entities: spatially fixed neurons and spatially propagating impulses. Neurons might be excitatory or inhibitory and their states are characterized by their subthreshold membrane potential or inner excitation, threshold level for firing, a resting level of inactivity state, maximum hyperpolarization level, absolute refractoriness period and a synaptic delay time. Under some conditions they emit impulses. Neurons are grouped in populations, state of the neurons in the population is described by the population's probability density function. Impulses move freely in space (in the numerical implementation some rule should be defined due to treat the effects due to spatial discretization), and might be absorbed by neurons chaining their inner excitation. Impulses are distributed in velocity–space according to the corresponding probability density function.

We extended [2, 16] this theory by using **diffusion theory** in two different senses. Both the dynamical behavior of neurons in their state-space and the movement of spikes in the physical space have been considered as diffusion processes. The state-space in the model consists of the two-dimensional space coordinate $\mathbf{r}$ for both neurons and spikes, a membrane potential coordinate u for all types of neurons, and an intracellular calcium-concentration coordinate χ for pyramidal neurons only. Both cell types, the inhibitory and excitatory ones are described by ionic conductances specific to neuronal type. Instead of fixed firing threshold a soft firing threshold is realized by voltage-dependent firing probability. Absorbed spikes induce time-dependent postsynaptic conductance change in neurons, expressed by the alpha-function.

We also realized in Budapest the importance of the existence of the database on connectivity in the cat cerebral cortex published in 1995, [36] and the necessity to include time delays. While our simulations of the activity propagation in hippocampus slices was based on the usual statistical assumptions, the first simulations by incorporating real connectivity data was done (well, with some time delay) by Tamás Kiss [23]. We (he) also took into account the axonal time delay. To calculate the synaptic current a new term $\varepsilon_s(\mathbf{r}, u, \chi, t)$ has been added:

$$\gamma_{s's} = \gamma_{s's}^{old} + \frac{\overline{\gamma'_{s's}}}{\tau'_{s's}} \int_0^\infty dt' \int_{\Omega(\mathbf{r}')} \kappa(\mathbf{r}, \mathbf{r}') \cdot a_{s's}(\mathbf{r}, t - t_d - t') \cdot t' \cdot \exp\left(1 - \frac{t'}{\tau'_{s's}}\right), \tag{13.1}$$

where the $\kappa(\mathbf{r}, \mathbf{r}')$ function determines the source and target cortical area between which information exchange occurs. Activity produced by the source population influences the target population after t_d time delay giving account of signal propagation delay in fibers.

The method and results were published in his master thesis written in Hungarian [23], (not necessarily the best marketing strategy). We made early not well-published (it is our fault) studies also on the disconnection syndromes, and simulated what it is now called *connectopathy* [8].

Statistical neurodynamics has at least two different features, as statistical mechanics. First, in neurodynamics "mean-field" approach is not enough, we should see both global and local dynamics. Our model gave the possibility to simulate the statistical behavior of large neural populations, and synchronously to monitor the behavior of an average single cell. Second, both statistical and specific cortical connections exist, model frameworks should describe their combination. In the project described in our last paper in the topic [24] both features were incorporated. As it was already written in [9], p. 272: "I think, each research group has bedroom secrets. The story with our "*population model*" is ours, and I think I should not blab it out."

Viktor Jirsa [20] classified the **mesoscopic models** into the following categories:

- Infinite Propagation Speed,Arbitrary Connectivity
- Finite Propagation Speed,Arbitrary Connectivity
- Infinite Propagation Speed, Symmetric, and Translationally Invariant Connectivity
- Infinite Propagation Speed, Symmetric and Translationally Variant Connectivity
- Finite Propagation Speed, Symmetric, and Translationally Invariant Connectivity
- Finite Propagation Speed,Asymmetric and Translationally Variant Connectivity

A general framework for neural field models with local and global connections also with time delay was given by him [21]. This model framework became the scientific basis of the Virtual Brain Project [45].

13.3 Forward and Inverse Modeling of the Neuro-Electric Phenomena

As we have seen in the previous paragraphs, that a strong branch of the modeling tradition in the neuroscience follows the bottom-up approach on the tracks of Hodgkin and Huxley. Starting from the biophysical mechanisms of the ion channels one can built neuron models on arbitrary levels of complexity. Then, connecting the neurons into networks, implementing connections from the basic synaptic dynamics up to the advanced activity dependent learning methods, one can study the emerging network dynamics. In the next step, as we will see here, solving the forward problem of the Poisson-equation, an artificial LFP can be synthesized and compared to the observed phenomena during electrophysiological measurements.

While the role of the modeling is less obvious in the top-down approach, it will be shown in this section, that models and modeling have an indispensable role, when we want to understand the measured electric signals by decomposing them into their sources. Here the model means a set of constrains and prior assumptions about the sources which implicitly or explicitly adopted by each method, to find a unique solution to source determination problem.

13.3.1 Micro-Electric Imaging

An average neuron in the cortex receives 10–15 thousand synapses from other neurons. While many fine details are known about the properties of individual synapses, and there is a progress on understanding brain connectivity [40], the spatio-temporal transmembrane current patterns, resulting from the summation of a huge number of individual synaptic inputs on a whole neuron, are almost entirely unknown. The main reason for this large gap in our knowledge is the lack of a proper technique for measuring spatio-temporal inputs patterns on single neurons in behaving animals. While the output of a neuron is well recognizable in the extracellular potential measurements in the form of action potentials, the input that evoked the observed spike is unknown. Without knowing the input, deciphering the input–output transformation implemented by an individual neuron is hopeless.

The steadily improving **optical imaging** techniques provide extremely good spatial resolution, but they still have not reached the speed, signal-to-noise ratio, sampling frequency, aperture and miniaturization properties necessary to record action potentials and synaptic input patterns on whole neurons in behaving animals.

On the other hand, the number of channels, together with the spatial resolution, have dramatically increased recently in the widely used multi-electrode arrays (MEA), and further improvements are expected [3, 6, 22]. This relatively low cost technique is applicable to freely behaving animals as well. Traditionally, only the spike timings are used from these extracellular (EC) potential recordings, but recent improvements significantly increased the spatial information content of these measurements. Thus, new techniques of data analysis are needed to exploit this new information.

We conclude that the rapid development of MEA techniques and the set of new analysis methods, directly designed to exploit the spatial information content of MEA recordings, may help to create a new emerging field to be called **micro-electric imaging**. Similar to the macroscopic imaging techniques, the different tasks of this field are: forward modeling, source reconstruction, anatomical area and layer determination, correlation and causality analysis while a specific task on this microscopic field is membrane potential and synaptic current reconstruction.

During the last few decades, a large variety of mathematical source reconstruction algorithms or imaging techniques have been developed for macroscopic neural electro-magnetic measurements, such as EEG and MEG. For a review, see [14]. However, on micro scales, only the traditional current source density (CSD) method has served the aim of identifying the neural transmembrane sources underlying the observed EC potential [30, 31]. The traditional CSD works well if the full 3-dimensional potential distribution is known with the spatial resolution comparable to the size of the sources. Definitely, current sources on single neurons cannot be analyzed this way, since 3D data cannot be collected by electrode systems without large tissue damage. Lacking this full 3D data, the CSD analysis based on 2D and 1D MEA measurements intrinsically requires the adoption of assumptions about homogeneity of the source density in the unknown dimensions. This homogeneity assumption can be a good approximation in the case of large population activities, but it is certainly not valid for single cell sources. Thus, we can conclude, that the (implicit) source model of the traditional 1D CSD analysis is an infinite homogeneous laminar source.

13.3.2 Source Reconstruction on Single Neurons

An alternative approach for CSD estimation is based on the inverse of the forward solution. To our knowledge, the first inverse CSD method was developed and applied to LFP data of olfactory bulb by Walter J. Freeman in 1980 [11, 12]. The inverse method was not applied to LFP data since, till it was rediscovered in recent years and applied to extracellular action potentials by [38] and local field potentials by [32].

The first inverse method for the estimation of cell-electrode distance and the reconstruction of the CSD on single neurons was introduced in our own lab [38] in 2005. The source model applied here called counter current model was a line-source, parallel to the electrode and consists of one high negative (sink) current peak on a smooth background of positive counter currents (sources). This model is valid only until the negative peak of the extracellular spike, so this single cell CSD method is able to calculate the CSD only at the peak of the action potential. Since then, numerous inverse CSD methods have been developed in many other research groups as well. Pettersen et al. [32] developed inverse CSD solutions for LFP, generated by a cortical column. The corresponding source model consists of homogeneous discs, whose laminar distribution was described either as sum of thin discs or a spline interpolated continuous distribution. Later, Daniel Wójcik and his group used kernel methods for 1, 2 and 3D inverse solutions [28, 29, 33]. The source models here consist of 3D Gaussian blobs and ensures a smooth inverse solution.

The recent sCSD method [39], built on the basis of the counter current model, is able to reconstruct the full **spatio-temporal CSD dynamics** of single neurons during the action potentials. By the sCSD method, the EC observability of back propagating action potentials in the basal dendrites of cortical neurons, the forward propagation preceding the action potential on the dendritic tree and the signs of the Ranvier-nodes has been demonstrated for the first time (Fig. 13.1).

13.3.3 Anatomical Area and Layer Determination: Micro-Electroanatomy

Proper interpretation of single neuron CSD maps during in vivo application of MEAs requires precise identification of the anatomical structures, cortical and synaptic layers in which the EC potentials were recorded. Post-hoc histology can provide information on the position of the probes in the brain, but it would be advantageous if this information would be accessible during the experiment as well, and in some

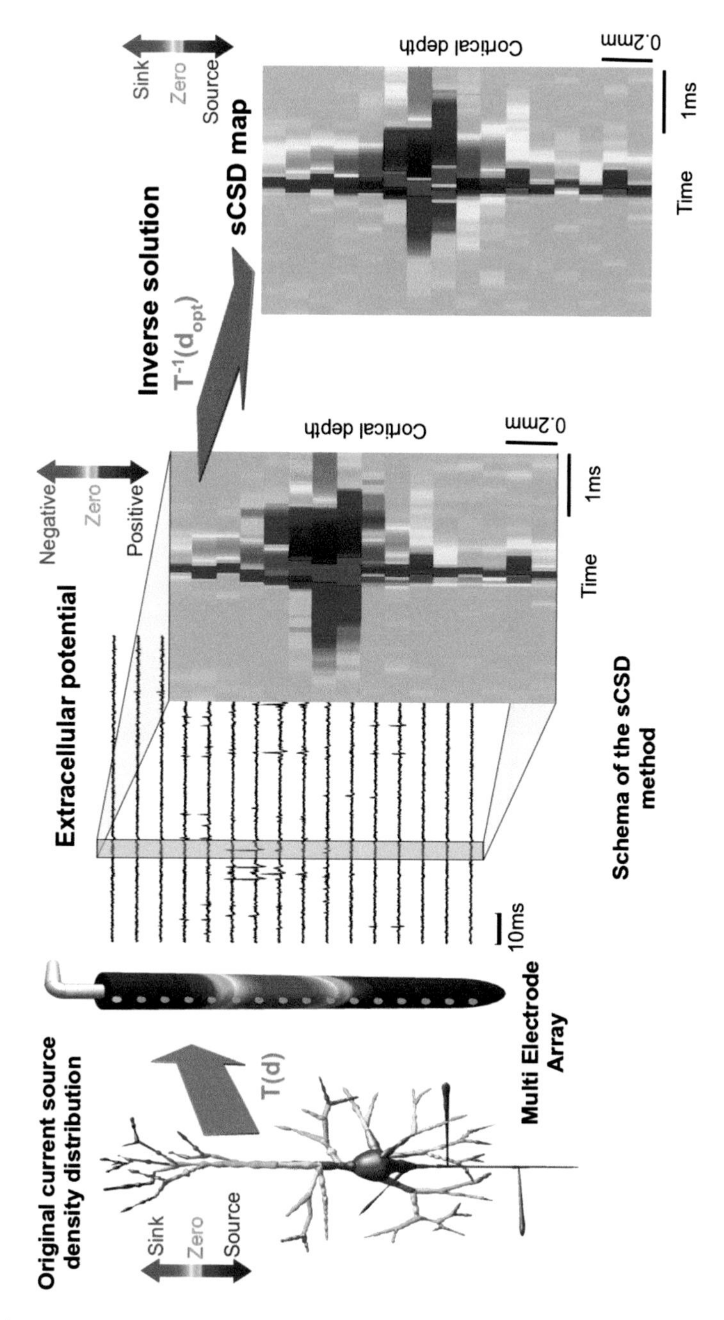

Fig. 13.1 The schema of the sCSD method: the *color-coded* current source density on the neuron determines the measured spatial potential pattern on the linear multi electrode array. This forward solution is expressed by the T(d) transfer matrix. The inverse solution, which we call sCSD method, starts from the measured EC potential of single neuron spikes. Then the application of the $T^{-1}(d)$ inverse transfer matrix yields the CSD distribution on the neuron. Modified from [39]

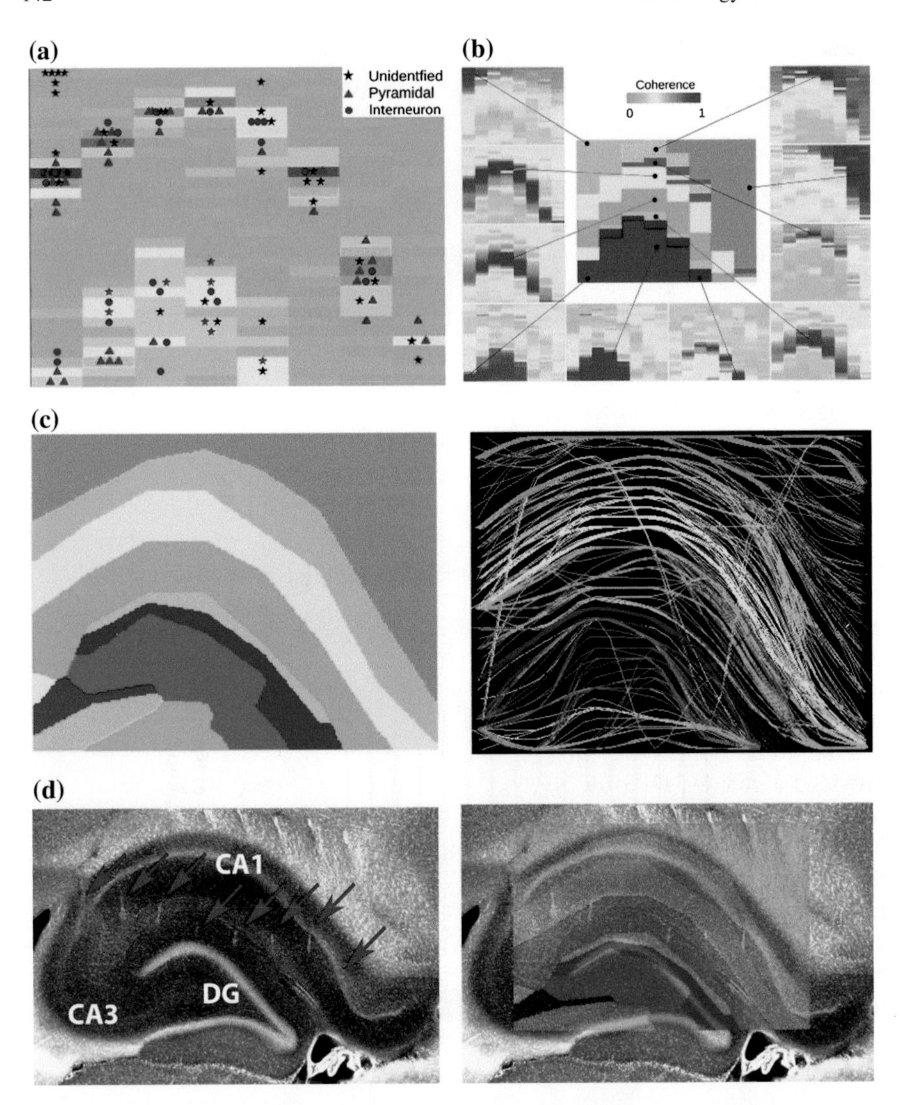

Fig. 13.2 Electroanatomy of the hippocampus. Somatic and synaptic layers are determined solely based on the recorded data. **a** Somatic layers were identified based on a high frequency (300 Hz) power map. **b** Synaptic layers were determined by coherence-based clustering. **c** The borders between layers and areas of the hippocampus is inferred by fusing the somatic map with the coherence-clusters. Our coherence-tracking algorithm visualizes the hippocampal anatomical structure clearly. **d** Comparison with histology. *Arrows* mark the paths of the 8 shanks of the electrode. From [3]

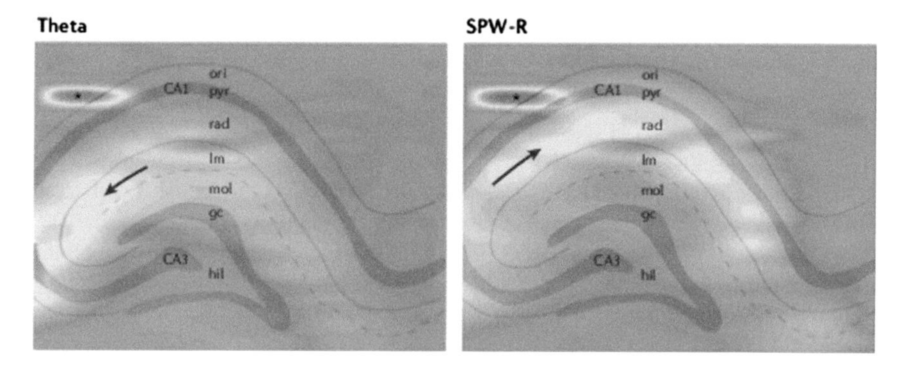

Fig. 13.3 Demonstration of different input patterns onto the same neuron. The same neuron (denoted by a *star*) is activated by different pathways and emits action potentials during theta and sharp-wave ripple oscillations. (Work of Z. Somogyvári and A. Berényi, from [7], Fig. 6)

cases the post-hoc histology cannot be performed well. The methodology of micro-electroanatomy [3], which was able to determine and visualize anatomical structures and synaptic layers in the hippocampus and in the neocortex solely based on the recorded multi-channel LFP data was a recent attempt on that. This anatomical reconstruction serves as a good basis for investigation of different synaptic input pathways on the neurons (Fig. 13.2).

Our preliminary results, previewed in the Nature Reviews Neuroscience [7] have provided a new insight into hippocampal dynamics, showing that the same CA1 interneuron receives input on different pathways in different hippocampal states. More precisely, the input was found to be dominated by the entorhinal perforant path during theta oscillations, but the Schaffer-collateral input from CA3 was stronger during sharp-wave ripple (SPW-R) periods. Thus, we conclude, that new, high-channel-count MEA data, precise identification of synaptic layers and model-based source reconstruction technique make possible a systematic analysis of synaptic input patterns for different cell types in different subregions of the hippocampus (Fig. 13.3).

13.4 Conclusions

We reviewed some specific concepts, where density functions play important role and may provide novel approaches for inferring, modeling and understanding neural dynamics and functions. In the first section we have briefly reviewed the application of density functions in the neural mass models as a 'golden midway' between to too detailed microscopic and the too phenomenological macroscopic approaches. This historical point of view led us to the conclusion, that the anatomical knowledge on the brain connectivity structure should be included into the pure statistical treatment as well. Besides some early attempts for our own laboratory, we recognized a strong trend into this direction in the recent years.

On the other side, density functions have inevitable role in the inference of the neural currents from the extracellular potential measurements, known as current source density analysis. In the CSD analysis, the collective effect of the abundant number of individual synapses is described by an appropriate density function. We have shown, that the solution of this inverse problem depends on the geometrical and dynamical assumptions about the sources. Different CSD methods use different source models, defining their range of validity and applicability. Finally we showed, that a new and promising branch of CSD methods emerged as the density functions have been applied to single neurons, allowing the inference of input current source density patterns on single neurons.

Acknowledgments ZS was supported by grant OTKA K 113147. PE thanks to the Henry Luce Foundation to let him to be a Henry R Luce Professor.

References

1. Amari S (1983) Field theory of self-organizing neural nets. IEEE Trans Syst Man Cybern SMC–13:741–748
2. Barna Gy, Grőbler T, Érdi P (1988) Statistical model of the Hippocampal CA3 region I. The single-cell module: bursting model of the pyramidal cell. Biol Cybern 79:301–308
3. Berényi A, Somogyvári Z, Nagy A, Roux L, Long J, Fujisawa S, Stark E, Leonardo A, Harris T, Buzski G (2014) Large-scale, high-density (up to 512 channels) recording of local circuits in behaving animals. J. Neurophysiol 111:1132–1149. doi:10.1152/jn.00785.2013BiolCybern, 79:309-321
4. Bergmann U, von der Malsburg C (2011) Self-organization of topographic bilinear networks for invariant recognition. Neural Comput 23:2770–2797
5. Bower JM, Beeman D (1994) The book of GENESIS: exploring realistic neural models with the GEneral NEural SImulation System. TELOS, Springer, New York
6. Buzsáki G (2004) Large-scale recording of neuronal ensembles. Nat Neurosci 7(5):446–51
7. Buzsáki G, Anastassiou CA, Koch C (2012) The origin of extracellular fields and currents–EEG, ECoG, LFP and spikes. Nat Rev Neurosci 13(6):407–420
8. Érdi P (2000) Narrowing the gap between neural models and brain imaging data: a mesoscopic approach to neural population dynamics. The 2000 Neuroscan Workshop at Duke University. http://www.rmki.kfki.hu/biofiz/cneuro/tutorials/duke/index.html
9. Érdi P (2007) Complexity explained. Springer, New York
10. Freeman WJ (1975) Mass action in the nervous system. Academic Press, Massachusetts
11. Freeman WJ (1980) A software lens for image reconstitution of the EEG. Prog Brain Res 54:123–127
12. Freeman WJ (1980) Use of spatial deconvolution to compensate for distortion of EEG by volume conduction. IEEE Trans Biomed Eng 27(8):421–429
13. Gerstner W, Kistler M, Naud R, Paninski (2014) Neuronal dynamics: from single neurons to networks and models of cognition. Cambridge University Press, Cambridge
14. Grech R, Cassar T, Muscat J, Camilleri KP, Fabri GS, Zervakis M, Xanthopoulos P, Sakkalis V, Vanrumste B (2008) Review on solving the inverse problem in EEG source analysis. J NeuroEng Rehabil 5(25):1–33
15. Griffith JA (1963) A field theory of neural nets. I. Derivation of field equations. Bull Math Biophys 25:111–120
16. Grőbler T, Barna Gy, Érdi P (1998) Statistical model of the Hippocampal CA3 region II. The population framework: model of rhythmic activity in the CA3 slice. Biol Cybern 79:309–321

17. Hines M (1984) Efficient computation of branched nerve equations. J Biol-Med Comp 15:69–74
18. Hines M (1993) The NEURON simulation program. Neural network simulation environments. Kluwer Academic Publication, Norwell
19. Hodgkin A, Huxley A (1952) A quantitative description of membrane current and its application to conduction and excitation in nerve. J Physiol 117:500–544
20. Jirsa VK (2004) Connectivity and dynamics of neural information processing. Neuroinformatics 2:183204
21. Jirsa V,K (2009) Neural field dynamics with local and global connectivity and time delay. Philos Trans R Soc A: Math Phys Eng Sci 367(1891):1131–1143
22. Kipke D, Shain W, Buzsáki G, Fetz E, Henderson J, Hetke J, Schalk G (2008) Advanced neurotechnologies for chronic neural interfaces: new horizons and clinical opportunities. J Neurosci 28(46):11830–11838
23. Kiss T (2000) Az agykéreg normális és epileptikus működésének tanulmányozása statisztikus neurodinamikai modellel (in Hungarian). Master's thesis, Eötvös Loránd Tudományegyetem. http://cneuro.rmki.kfki.hu/files/diploma.pdf
24. Kiss T, Érdi P (2002) Mesoscopic Neurodynamics. BioSystems, Michael Conrad's special issue 64(1–3):119–126
25. Kozma R, Freeman WJ (2003) Basic principles of the KIV model and its application to the navigation problem. J Integr Neurosci 2(1):125–145
26. Kozma R, Freeman WJ, Érdi P (2003) The KIV model—nonlinear spatio-temporal dynamics of the primordial vertberate forebrain. Neurocomputing 52–54:819–826
27. Kozma R, Freeman WJ, Wong D, Érdi P (2004) Learning environmental clues in the KIV model of the Cortico-Hippocampal formation. Neurocomputing 58–60(2004):721–728
28. Leski S, Wajcik DK, Tereszczuk J, Awiejkowski DA, Kublik E, Wrabel A (2007) Inverse Current-Source Density in three dimensions. Neuroinformatics 5:207
29. Leski S, Pettersen KH, Tunstall B, Einevoll GT, Gigg J, Wajcik DK (2011) Inverse Current Source Density method in two dimensions: inferring neural activation from multielectrode recordings. Neuroinformatics 9:401–425
30. Mitzdorf U (1985) Current source-density method and application in cat cerebral cortex: investigation of ecoked potentials and EEG phenomena. Physiol Rev 65:37–100
31. Nicholson C, Freeman JA (1975) Theory of current source-density analysis and determination of conductivity tensor for anuran cerebellum. J Neurophysiol 38:356–368
32. Pettersen KH, Devor A, Ulbert I, Dale AM, Einevoll GT (2006) Current-source density estimation based on inversion of electrostatic forward solution: effect of finite extent of neuronal activity and conductivity discontinuites. J Neurosci Methods 154(1–2):116–133
33. Potworowski J, Jakuczun W, ęski S, Wjcik DK (2012) Kernel current source density method. Neural Comput 24:541–575
34. Rall W (1962) Electrophysiology of a dendritic neuron model. Biophys J 2:145–167
35. Rall W (1977) Core conductor theory ad cable properties of neurons. Handbook of physiology. The nervous system. William and Wilkins, Baltimore, pp 39–98
36. Scannell JW, Blakemore C, Young MP (1995) J Neurosci 15:1463–1483
37. Seelen W (1968) Informationsverarbeitung in homogenen netzen von neuronenmodellen. Kybernetik 5:181–194
38. Somogyvári Z, Zalányi L, Ulbert I, Érdi P (2005) Model-based source localization of extracellular action potentials. J Neurosci Methods 147:126–137
39. Somogyvári Z, Cserpán D, Ulbert I, Érdi P (2012) Localization of single-cell current sources based on extracellular potential patterns: the spike CSD method. Eur J Neurosci 36(10):3299–313
40. Sporns O (2010) Networks of the brain. MIT Press, Cambridge
41. Ventriglia F (1974) Kinetic approach to neural systems. Bull Math Biol 36:534–544
42. Ventriglia F (1982) Kinetic theory of neural systems: memory effects. In: Trappl R (ed) Proceedings of the Sixth European Meeting on Cybernetics and Systems Research. Austrian Society for Cybernetic Studies, North-Holland Publishing Company, Amsterdam, pp 271–276

43. Ventriglia F (1990) Activity in cortical-like neural systems: short-range effects and attention phenomena. Bull Math Biol 52:397–429
44. Ventriglia F (1994) Towards a kinetic theory of cortical-like neural fields. Neural modeling and neural networks. Pergamon Press, Oxford, pp 217–249
45. The Virtual Brain Project. http://www.thevirtualbrain.org/tvb/zwei
46. Wilson HR, Cowan J (1973) A mathematical theory of the functional dynamics of cortical and thalamic neurons tissue. Kybernetik 13:55–80
47. Willshaw DJ, von der Malsburg C (1976) How patterned neural connections can be set up by self-organization. Proc R Soc Lond B194:431–445

Chapter 14
Commentary by Frank Ohl

On the Creation of Meaning in the Brain—Cortical Neurodynamics During Category Learning

Frank W. Ohl

14.1 Introduction

Large-scale spatiotemporal activity patterns in neocortex have been observed since the first attempts of simultaneous multichannel recording (e.g. [20–22]) and are now regularly reported by studies using appropriate techniques on both microscopic [12] and macroscopic spatial scales [10]. Previous work, predominantly on the spatiotemporal activity patterns in the olfactory bulb and cortex of rabbits trained to discriminate odors [8] has suggested (1) that spatiotemporal activity patterns on a mesoscopic scale are not merely an epiphenomenon of an underlying activity in a complex neuronal network, but represent a spatially extended manifestation of large-scale neuronal coordination, and (2) that this field-like organization of coordinated activity is associated with endogenous cognitive states mediating the meaning associated with received stimuli, rather than simply representing the "arrival" of a stimulus or its physical parameters.

Here we review a set of studies that have recently improved our understanding of the nature of large-scale coordinated activity on a mesoscopic scale (see also [25]) by exploiting two experimental conditions. The first condition is that the measurements were taken not from the olfactory system but from the auditory cortex which has a very strict topographic representation of some physical stimulus features (most prevalent is the tonotopic representation of dominant spectral content of sound) unlike the olfactory cortex. It was hypothesized that this condition would facilitate the experimental dissociation between topographic and field-like holographic organizational principles of spatially extended activity patterns. Here we use the terms topographic and holographic to emphasize a difference in the spatial organization of information about a stimulus and its meaning. It should be noted that biophysical mechanisms reminiscent of optical holography underlying cortical physiology

F.W. Ohl
Leibniz Institute for Neurobiology, Magdeburg, Germany

R. Kozma and W.J. Freeman, *Cognitive Phase Transitions in the Cerebral Cortex – Enhancing the Neuron Doctrine by Modeling Neural Fields*,
Studies in Systems, Decision and Control 39, DOI 10.1007/978-3-319-24406-8_14

have been proposed previously already for the context of stimulus processing alone [34, 35]. The second experimental condition is that instead of discrimination learning, category formation was invoked. Unlike discrimination learning, category formation involves the establishment of equivalence classes of meaning despite variance in physical stimulus parameters, thereby allowing a dissociation of feature representation and meaning representation in an experimentally controlled way.

In the following sections we first describe why auditory cortex is particularly suited to design a study that critically tests the implications of the hypothesis that large-scale cortical activity patterns are associated with the semantics of sensory stimuli. We then describe the logic of the category formation paradigm used to dissociate feature representation from meaning representation. Finally we present the evidence that the latter is associated with holographically organized field-like activity patterns observable in single trials after stimulus presentation that can coexist in the same neuronal tissue with a topographically organized, input-driven representational principle that is traditionally known in sensory neocortices and can be observed even after (and, in fact, particularly clearly after) averaging over multiple stimulation trials.

14.2 Traditional Conceptualizations of Auditory Cortex

The traditional conceptualization of auditory cortex has been that of "processing module" in a hierarchical series of interconnected modules, functionally dedicated to the analysis of sound. Analogous conceptualizations do exist for the sensory cortices of other sensory modalities. The envisaged processing tasks of these modules have primarily been associated with the concept of a "flow of neuronal excitation" through a series of processing stations, from the sensory epithelium excited by a sensory stimulus along an "ascending pathway" towards the sensory cortex as the "top module" or at least a module relatively late in the processing sequence for sensory stimuli.

In this conceptualization, "analysis of sound" pertains to the extraction, representation and computation of physical features characterizing the sound or the inferred sound source. Consequently, the search for neuronal mechanisms putatively serving these three functions has been the hallmark of traditional research on the physiology of auditory cortex. Typical findings of this approach have been (1) the identification spectral or spectro-temporal filter properties in single neurons across the entire auditory system, indicative of the "extraction" and "representation" of sound frequency, or more complex physical sound features, or (2) the "computation" of derived features, for example the sound source localization based on the neural representation of interaural cues. Notably, for this approach typical descriptors of neuronal activity on both the micro-scopic level (e.g. spectrotemporal receptive fields) and macroscopic level (e.g. event-related potentials) are based on averaging over multiple stimulus presentations. In case of unfavorable statistics of the stimuli [23] or stimulus-evoked neuronal activities [24] simple averaging approaches cannot be used.

This approach has been highly efficient in unraveling several important functional principles of the auditory system organization (for overview see [42]). Problems of this approach persist for example in cases where nonlinearities of the input-output behavior of single neurons or of entire auditory nuclei impair identification of the "computational operations" implemented, or where lack of knowledge about the functional connectivity impairs identification of the relevant network of neurons or nuclei involved. Among the consequences of these persistent problems are (1) a certain arbitrariness in the selection of perception-relevant auditory sound features for which a "neuronal processing" has so far been identified and (2) the fact that for many aspects of auditory processing the neuronal basis has remained elusive.

14.3 Learning-Induced Plasticity in Auditory Cortex and Multisensory Processing

The traditional conceptualization of auditory cortex as a mere stimulus analyzer has been challenged first by demonstrating that learning experiences of an animal subject in conditioning paradigms can significantly alter the response patterns of single neurons [39] or cortical masses [11] to auditory stimuli, despite the constancy of the physical stimulus parameters. For example, in classical conditioning paradigms using pure tones as the conditioned stimulus, conditioning shifted the preferred frequency of single neurons' receptive fields from their original value towards the frequency of the conditioning tone [38]. In experimental paradigms training the discrimination between neighboring frequencies, local increases of the tuning slope in the spectral neighborhood of the discriminated frequencies were observed [26, 27]. Today, numerous studies have demonstrated phenomena of learning-induced plasticity in the auditory cortex and responses of neurons in auditory cortex to auditory stimuli are generally considered to be recruited in a highly task-dependent manner (for review see [6, 28, 36, 37]).

Independent evidence indicating the necessity for reconceptualizing auditory cortex function came from studies of multisensory processing. Apparently, neuronal activity in the auditory cortex can be modulated [18, 19] or even evoked [4] by stimuli from other sensory modalities, or from non-auditory stimuli that have previously been associated with auditory stimuli signaling behaviorally relevant events [5]. Consequently, auditory cortex is no longer viewed as merely the top module in a hierarchy of processing modules but as a highly plastic system embedded in a network of bottom-up and top-down processes [25].

Additional independent evidence that cortical processing in general might not be well characterized by concepts of extracting "features" from a stimulus, or re-composing a sensory object from its previously isolated "features" is provided by models of cortical development and sensory processing that derive many aspects of cortical known physiology from a few fundamental assumptions about the dominating role of competition for limited metabolic resources among developing cortical neurons [43].

14.4 Towards Understanding the Neurodynamics Underlying Perception and Cognition

Studies of the functional role of large-scale neurodynamics were pioneered primarily in the olfactory bulb and cortex [8] and have demonstrated the propaedeutic relevance of the electro-physiological recording of field-like neuronal mass action on a mesoscopic spatial scale in addition to microscopic (single-neuron) activity [10]. After early studies had indicated that spontaneous and stimulus-triggered cortical activity are organized in large-scale spatiotemporal activity patterns [20–22] studies in the olfactory bulb and cortex of rabbits trained to discriminate odors showed that these patterns systematically varied with learning. However these patterns could not be simply interpreted as stimulus representations as they changed for an otherwise constant stimulus with current context and previous stimulation history. It was therefore concluded that stimulus-associated cortical spatiotemporal activity patterns should not be considered "stimulus representations" but dynamical activity states reflecting endogenous cognitive processes in the animal subjects.

14.5 Exploiting Category Formation to Study the Neurodyamics Underlying the "Creation of Meaning" in the Brain

The above conclusion is clearly compatible with the demonstrated lack of invariance of cortical spatiotemporal activity patterns with unchanging stimuli. However, we have argued that a more direct evidence for this conclusion would additionally require demonstration of pattern invariance in an endogenous cognitive process during which subjective invariance is generated despite variance in the physical stimulus characteristics. It has been a long-standing theorem in the cognitive sciences that the formation of categories is exactly this process [15]. Category formation enables the establishment of cognitive sharp boundaries in physical continua and the collection of heterogeneous objects into a unified set. Categories and category boundaries can exist as inborn cognitive structures or can be established based on learning. In any case, categories are equivalence classes of meaning [7] and are therefore an ideal instrument to study the biophysical processes underlying the emergence of subjective cognitive structures.

Specifically we hypothesized that if cortical spatiotemporal activity patterns indeed reflect endogenous subjective aspects of perception then the formation of categories should introduce a detectable invariance in these patters at the onset of categorical grouping. To test this hypothesis we have developed a rodent model (Mongolian gerbil, *Meriones unguiculatus*) of auditory category learning that employed linearly frequency modulated tones as stimuli [32, 40]. Such stimuli are suited to study stimulus-induced cortical activity, because they lead to large-scale activity patterns [31]. More importantly, discrimination of the modulation direction of frequency-modulated tones traversing the same frequency range in opposite directions was shown by lesion studies to depend on auditory cortex, unlike discrimination of many other auditory cues [29, 41].

Here we refer to a set of experiments in which gerbils were trained to categorize the modulation direction ("rising" or "falling") of frequency-modulated tones. Training was organized as a series of subsequent training blocks. In a first training block the discrimination of two frequency-modulated tones traversing the same frequency interval (e.g. from 2 to 4 kHz) with rising or falling instantaneous frequency, respectively, was trained. For training and psychophysical assessment of the discrimination performance, a go/(no go) paradigm was used employing a two-way shuttle-box active avoidance procedure. In such a training block a more or less gradual increase of the behavioral hit rate up to an asymptotic value can be typically observed (Fig. 14.1a). In the asymptotic phase a stable psychometric function can be measured upon variation of physical stimulus parameters, like for example the modulation rate of the instantaneous frequency of the tone. The psychometric function is typically characterized by a gradual fall-off of a suitable measure of stimulus-specific response behavior (often strength of a response, or frequency of occurrence of a response) with distance of a characteristic physical parameter from the trained value (Fig. 14.1c).

The existence of non-vanishing responses to stimuli other than the trained stimulus is called generalization, not to be confused with categorization (see below), the gradual fall-off in the psychometric function is called the generalization gradient. When a so-trained animal is subsequently trained in a second training block with a new pair of frequency-modulated tones, and these tones are so selected that their parameters fall well outside the previously established generalization gradients, a similar learning curve, starting with non-significant performance values and later reaching an asymptote, can be observed. If several such training blocks are trained after another, in each of which (say) the rising tone must be responded to with a go response and the falling tone must be responded to with a no-go response, a behavioral state transition can be typically observed after a number of training blocks. This transition is characterized by the correct association of the novel stimuli in a newly started training block with the corresponding go and no-go behaviors already from the first training session on, giving rise to a learning curve as depicted in Fig. 14.1b. In that cognitive state the psychometric function attains a sigmoid shape indicating a parcellation of the stimulus parameter space into different categories separated by a sharp boundary (categorization boundary), Fig. 14.1d. Hence, while generalization is a hardly avoidable phenomenon accompanying simple stimulus detection and discrimination, categorization is an active cognitive process based on the semantic parcellation of the represented world into equivalence classes of meaning. The cognitive states associated with simple discrimination or with categorization, can thus be dissociated by their learning dynamics (Fig. 14.1a, b) or by the corresponding psychometric functions (Fig. 14.1c, d).

The cognitive transition from discrimination to categorization can be emphasized by analyzing the difference between hit and false alarm rate measured at the beginning of a training block (e.g. during the first training session). Figure 14.2a depicts

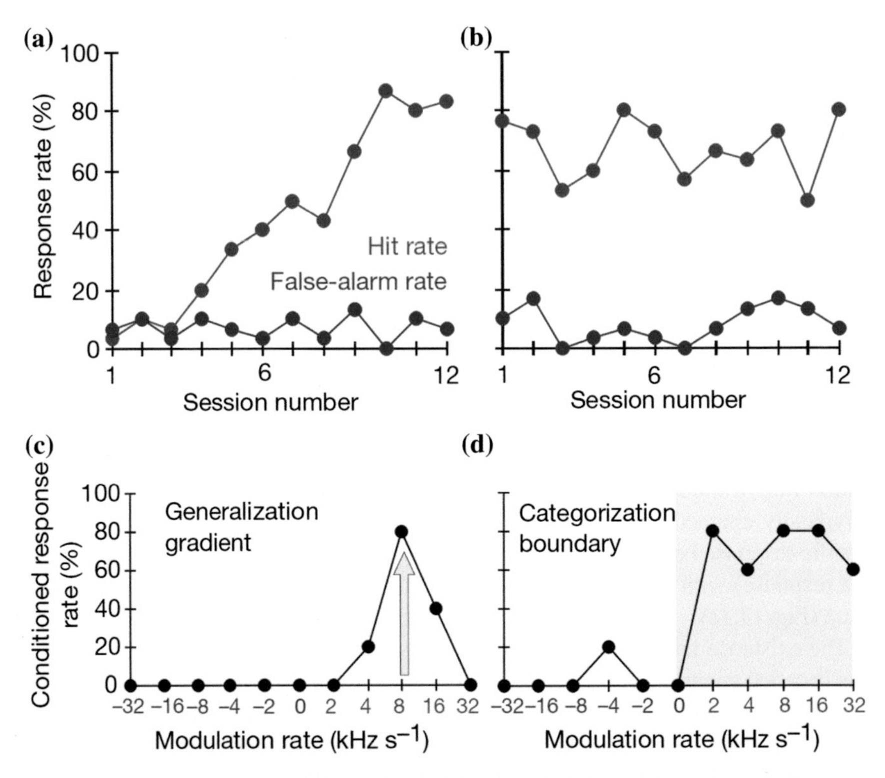

Fig. 14.1 Representative data of an individual learning curve and psychometric function before and after the transition from discrimination behavior to categorization behavior. *Panels a* and *b* display the learning curves before and after the transition respectively. Note the gradual increase of the hit rate with training sessions index during the discrimination phase and the high discrimination performance already at the beginning of the training in the categorization phase. *Panels c* and *d* show the psychometric functions for variations of the rate of modulation of instantaneous frequency of the FM tomes. Note the sigmoid shape ("curve of categorical perception") of the psychometric function after the transition. From [32]

this difference for a representative individual as a function of training block index; apparently the transition from discrimination to categorization occurred between the fourth and fifth training block. The cognitive state transitions were characterized by (1) the abruptness (rather than gradual development) of the transition, (2) the individuality of the point in time of its occurrence in each animal subject, and (3) the stability of the categorization after the transition. These are three key features of what in cognitive science is sometimes termed an "*Aha*" event, a sudden insight changing the cognitive structure of an individual.

Analysis of cortical spatiotemporal activity patterns recorded with epidurally implanted multielectrode arrays allowed us to study the variance of stimulus-associated patterns in different phases of the learning history of each individual. Patterns recorded with arrays of N electrodes were described by N-dimensional vec-

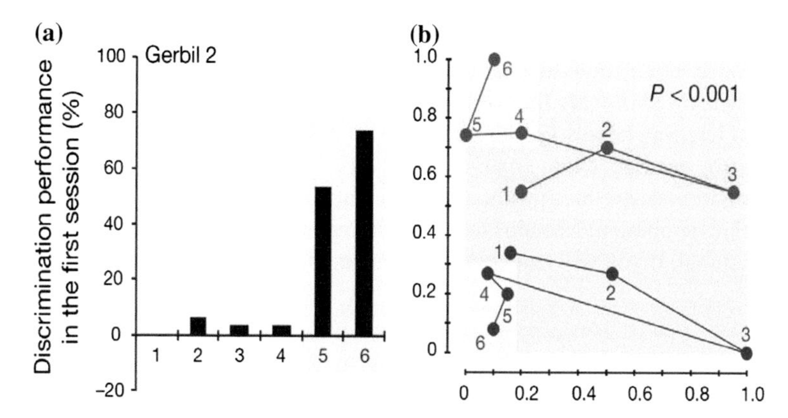

Fig. 14.2 Corresponding psychometric and electrophysiological data of the "*Aha*" event during category formation in one representative individual. **a** Discrimination performance measured by the difference between hit and false alarm rate in the first training session of six subsequent training blocks (abscissa). Note the abrupt onset of categorization with the 5th training block. **b** Graphical display of the similarity relations between mesoscopic cortical activity patterns associated with stimuli from the two categories (*red*: rising FM, *blue*: falling FM) during different training blocks (numbers) obtained by nonlinear projection of pattern vectors from an 18-dimensional space into two dimensions. Note that with the onset of categorization (training block 5), projected vectors stop moving away from the previously attained vector (index 4) forming clusters in the state space. From [32]

tors, each vector component representing the averaged signal power in small time window (for details see [32]). As was shown in previous studies (e.g. [1]), these vectors allowed robust classification of the recorded cortical activity with respect to the stimulus employed in a given trial. When the classification performance of these vectors was studied as a function of time, several transient local maxima occurred in the time (typically 4 s) between stimulus presentation and behavioral response. A first maximum occurred as expected with a few milliseconds delay after the stimulus onset ("early pattern"), but additional transient maxima with variable latencies occurred especially with training ("late patterns") [32]. While the early patterns were trivially expected as they represented the activity associated with the afferent volley of excitation relayed from the thalamus to the cortex, we focused for the further analysis on the late patterns which we called "marked states". For convenient visual display of the statistical analyses of the multivariate data, we projected the patterns associated with different stimuli and different learning phases into 2-dimensional space using a non-linear mapping algorithm that keeps all pairwise distances between vectors in N-space approximately constant (for details see [32]). Figure 14.2b shows the result for the individual animal whose behavioral data are depicted in Fig. 14.2a. It is obvious that the dissimilarity (indicated by distances between projected points) between patterns belonging to the same category (red: rising, blue: falling) is of the same order of magnitude than between categories during the first four training blocks when the animal is still in its discrimination phase. After the transition to the categorization phase (training blocks 5 and 6) the patterns remain in the vicinity of the previously attained giving rise to "clusters" in the projected space. This finding, which was verified in every individual (although happening at different times in each individual learning history) confirmed the hypothesis of a physiological correlate of the cognitive invariance that emerged at the onset of categorization. It also showed that these patterns have metric properties, but that these can no longer be explained by topographic mapping of stimulus features but must be the result of an organizational (computational, if you will) process reflecting an individual's psychometric scaling.

14.6 Coexistence of Point-Like Topographic and Field-Like Holographic Representation of Information

The above described data already indicated a different nature of the early patterns that occurred in a time-locked fashion with fixed short latency after the stimulus and the late patterns that emerged from the ongoing activity at variable times. To obtain insight in the spatial organization of information in these patterns about the stimuli and categories we performed the following two-step analysis.

In a first step we randomly removed a number n_r of electrode channels (vector components) from the original pattern (containing 18 channels) and studied the decline of discrimination performance with respect to stimuli used in the trials. For each number of electrode channels to be removed, the removal was repeated multiple times with a randomly selected set of n_r channels (for details see [33]). Figure 14.3 depicts, separately for early and late patterns, the maximum number, minimum number and median number of correct classifications in the 60 trials of a representative training session taken from the asymptote of the learning curve, as a function of the number of remaining electrodes (vector components) in the pattern. Since the set of removed channels was randomly selected, the median values represent a measure of the "density" of channels required to capture the discriminant information about the two stimuli. The median values declined with decreasing number of remaining channels in a gradual manner for the early patterns and in a more sigmoid pattern for the late patterns. These different behaviors of the density measure can be understood by investigating the behavior of the minimum and maximum number of correct classifications: The maximum number showed a sigmoid dependence from the number of remaining electrodes in both early and later patterns, but only for early patterns the minimum number was found to remain on chance level, even for high number of remaining electrodes. This indicated (1) that in both early and late patterns high discrimination performance was possible even with a reduced set of channels (down to 11 remaining electrodes), but (2) that only in early patterns it was possible to completely destroy stimulus- or category-specific information by removing only a small number (e.g. 1 or 2) of channels (corresponding to 16 or 17 remaining channels). This analysis therefore provided first evidence for focal or

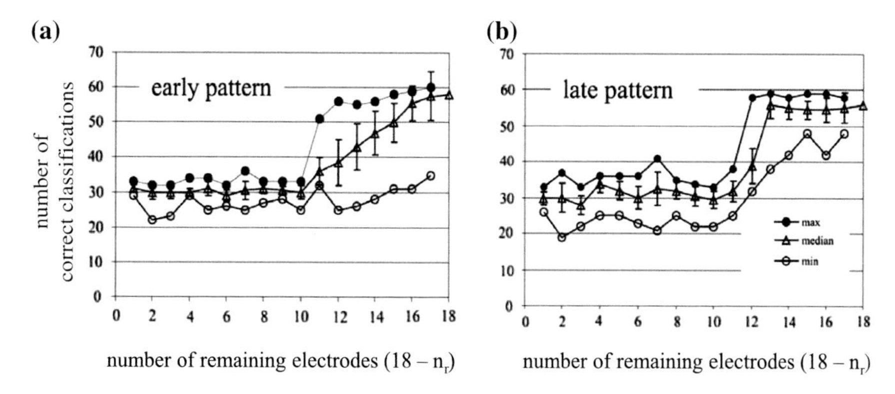

Fig. 14.3 Number of correct classifications of early (**a**) and late patterns (**b**) in 60 trials of a training session when the animal had reached the plateau phase of its learning curve. Shown are the minimum, median and maximum number of correct classifications ($\pm SEM$) after 40 (2–16 remaining electrodes) or 18 (1–17 remaining electrodes) repetitions of removing n_r randomly selected channels from the pattern. The triangles without error bars above the abscissas value of 18 correspond to the number of correct classifications observed in the experiment. Note the small values for the minimum number of correct classifications in the early but not in the late pattern. From [33]

point-like spatial organization of discriminant information in the early patterns as opposed to a field-like, spatially distributed organization of discriminant information in the late patterns.

In a second step we estimated how stimulus-discriminating information was spatially organized in early and late patterns. To this aim we determined the sets of most discriminating channels in early and late patterns, i.e. the sets of channels the removal of which from the pattern caused the largest decline in classification performance by the pattern. Figure 14.4 depicts the sets of most discriminating channels marked in orange on the spatial array of recording channels. Shown are the sets of most discriminating channels for $n_r = 2$, 4, 11, and 12 removed channels in early and late patterns for four animals (rows). The inset in the upper left is given for spatial reference and shows the 50 % isopotential contours of the early evoked activity (P1 component of the middle latency potential) for the rising and falling FM tone used in this session, together with an anatomical landmark. The 50 % isopotential contours reflect the spatial organization of the dominant thalamocortical input into the cortex [2, 3, 30]. This physiological observable is a hallmark of the topographic aspects of cortical organization: variation of the frequency content of the auditory stimulus leads to a topographic shift of this contour along the tonotopic axis in primary auditory cortex field A1 [30]. Also, the fact that frequency-modulated tones traversing the same frequency interval in different modulation directions lead to contours that are largely overlapping by slightly shifted against each other in the direction of the tonotopic gradient was demonstrated to result from the tonotopic organization of thalamocortical input into the cortex [31].

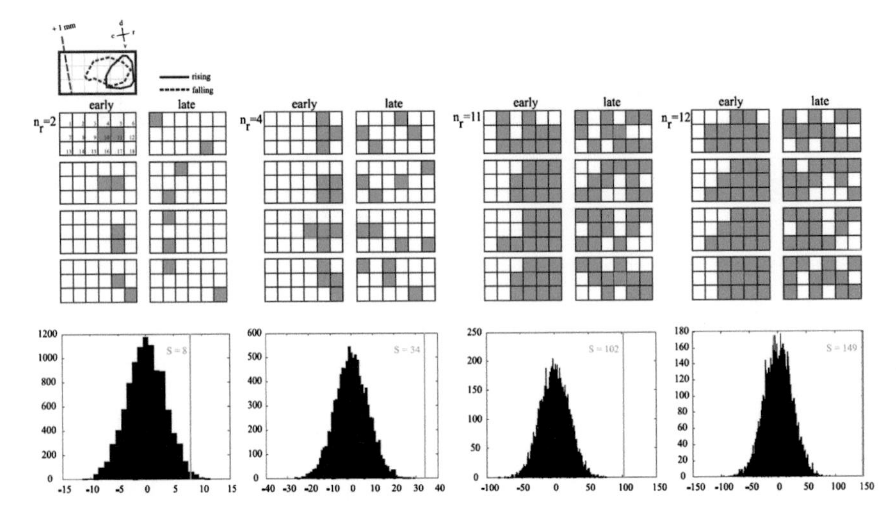

Fig. 14.4 Spatial organization of the sets of most discriminating channels. The inset at the *top left* shows the topography of the 50 (%) isopotential contour of the P1 component of the event-related potential for rising and a symmetrically falling FM tone. The diagrams below show the n_r most discriminating electrodes in early and late patterns for 4 animals (rows) and selected values of n_r (2, 4, 11, and 12). Note the clustered organization of each set of most discriminating electrodes in the early patterns and the spatially distributed organization in the late patterns. The histograms show the test statistic S (see text) and its distribution after resampling (10000 times) the interelectrode distances within the sets of most discriminating electrodes

The spatial distributions of most discriminating electrodes were found to be markedly different in the early and late patterns. It can be seen that for the early patterns the sets of most discriminating channels tend to be connected sets (corresponding electrodes are neighbors of each other) and co-localized with the dominant thalamic input (see inset). For the late patterns, the most discriminating electrodes are not connected but form sets spatially distributed over the entire recording array. To quantify this difference in spatial organization between early and late patterns we calculated for each pattern the distances between each pair of most discriminating electrodes contained in a set based on the Moore neighborhood, i.e. channels adjacent in the horizontal, vertical or diagonal direction were considered to have a distance of 1 to each other. We calculated a test statistic $S = L - E$, in which E and L are the sums over all distances between most discriminating electrodes in the early and late patterns, respectively, across all animals. A higher compactness of the set of most discriminating electrodes in the early patterns in comparison to late patterns should yield $S > 0$. The significance of the empirically found S values was calculated with respect to a distribution of S values obtained with 10,000 randomly chosen distance values experimentally observed for a given n_r (Fig. 14.4, bottom row). Our data indicate a predominantly topographical organization of the discriminant information for the early patterns and a predominantly holographical organization for the late patterns. Apparently, both types of organization co-exist in the same neural tissue.

14.7 Conclusion and Outlook

We have reviewed behavioral/psychophysical and electrophysiological evidence from a rodent model of auditory category learning for the hypothesis that stimulus-associated mesoscopic field-like activity patterns in auditory cortex are associated with the semantics of sensory stimulus processing. In these patterns stimulus-discriminating information is holographically organized, i.e. extended spatially over the entire recording region with no focal concentration of local information. These patterns coexist in the same neuronal tissue with traditionally known patterns that are strictly stimulus-locked and can therefore be seen after averaging across multiple stimulus presentation. In these early patterns stimulus-discriminating information is topographically organized.

Among the next challenges will be the understanding of how these patterns can develop from the ongoing dynamics in cortical networks, how this development depends on learning and other forms of plasticity, and what role such patterns, once established, play for the future cognitive processing of an individual. It must be realized that if pattern formation depends on cooperative mechanisms involving very large numbers of neurons, a mesoscopic abstraction from microscopic (single-unit) data must be constructed, similar to other areas of mesoscopic physics. A better understanding of the relationships between microscopic and mesoscopic levels of neuronal activity will therefore be necessary. Two central aspects will likely be the elucidation of destabilizing effect that sensory stimuli exert on the ongoing dynamics [9] and the phase-transition dynamics of ongoing and stimulus-evoked activity [16, 17]. Recently established experimental methods to separate thalamocortical ("input-driven") from cortico-cortical ("endogenous") contributions to cortical mass activity [13, 14] might be instrumental to design experimentally testable models of mesoscopic cortical dynamics.

References

1. Barrie JM, Freeman WJ, Lenhart MD (1996) Spatiotemporal analysis of prepyriform, visual, auditory, and somesthetic surface EEGs in trained rabbits. J Neurophysiol 76:520–539
2. Barth DS, Di S (1990) Three-dimensional analysis of auditory-evoked potentials in rat neocortex. J Neurophysiol 64:1527–1536
3. Barth DS, Di S (1991) The functional anatomy of middle latency auditory evoked potentials. Brain Res 565:109–115
4. Brosch M, Selezneva E, Scheich H (2005) Nonauditory events of a behavioral procedure activate auditory cortex of highly trained monkeys. J Neurosci 25:6797–6806
5. Cahill L, Ohl F, Scheich H (1996) Alteration of auditory cortex activity with a visual stimulus through conditioning: a 2-deoxyglucose analysis. Neurobiol Learn Mem 65:213–222
6. Dahmen JC, King AJ (2007) Learning to hear: plasticity of auditory cortical processing. Curr Opin Neurobiol 17:456–464
7. Estes WK (1994) Classification and cognition. Oxford University Press, Oxford
8. Freeman WJ (1975) Mass action in the nervous system. Academic Press, New York
9. Freeman WJ (1994) Neural mechanisms underlying destabilization of cortex by sensory input. Phys D 75:151–164

10. Freeman WJ, Quiroga RQ (2013) Imaging brain function with EEG. Advanced temporal and spatial analysis of electroencephalographic signals. Springer, New York
11. Gonzalez-Lima F, Scheich H (1984) Neural substrates for tone-conditioned bradycardia demonstrated with 2-deoxyglucose. I. Activation of auditory nuclei. Behav Brain Res 14:213–233
12. Grün S, Rotter S (eds) (2010) Analysis of parallel spike trains. Springer, New York
13. Happel MF, Jeschke M, Ohl FW (2010) Spectral integration in primary auditory cortex attributable to temporally precise convergence of thalamocortical and intracortical input. J Neurosci 30:11114–11127
14. Happel MF, Deliano M, Handschuh J, Ohl FW (2014) Dopamine-modulated recurrent corticoefferent feedback in primary sensory cortex promotes detection of behaviorally relevant stimuli. J Neurosci 34:1234–1247
15. Kant I (1787) Kritik der reinen Vernunft. Universitt Duisberg-Essen
16. Kozma R, Puljic M, Balister P, Bollobs B, Freeman WJ (2005) Phase transitions in the neuropercolation model of neural populations with mixed local and non-local interactions. Biol Cybern 92:367–379
17. Kozma R, Puljic M, Freeman WJ (2012) Thermodynamic model of criticality in the cortex based on EEG/ECoG data. In: Plenz D (ed) Criticality in neural systems. Wiley, New York, pp 153–176
18. Lakatos P, Chen CM, O'Connell MN, Mills A, Schroeder CE (2007) Neuronal oscillations and multisensory interaction in primary auditory cortex. Neuron 53:279–292
19. Lakatos P, O'Connell MN, Barczak A, Mills A, Javitt DC, Schroeder CE (2009) The leading sense: supramodal control of neuro-physiological context by attention. Neuron 64:419–430
20. Lilly JC (1954) Instantaneous relations between the activities of closely spaced zones on the cerebral cortex; electrical figures during responses and spontaneous activity. Am J Physiol 176:493–504
21. Lilly JC, Cherry RB (1954) Surface movements of click responses from acoustic cerebral cortex of cat: leading and trailing edges of a response figure. J Neurophysiol 17:521–532
22. Lilly JC, Cherry RB (1954) Surface movements of figures in spontaneous activity of anesthetized cerebral cortex: leading and trailing edges. J Neurophysiol 18:18–32
23. Meyer AF, Diepenbrock JP, Happel MF, Ohl FW, Anemller J (2014) Discriminative learning of receptive fields from responses to non-Gaussian stimulus ensembles. PLoS ONE 9:e93062
24. Meyer AF, Diepenbrock JP, Ohl FW, Anemller J (2014) Temporal variability of spectro-temporal receptive fields in the anesthetized auditory cortex. Front Comput Neurosci 8:165
25. Ohl FW (2014) Role of cortical neurodynamics for understanding the neural basis of motivated behavior—lessons from auditory category learning. Curr Opin Neurobiol 31C:88–94
26. Ohl FW, Scheich H (1996) Differential frequency conditioning enhances spectral contrast sensitivity of units in auditory cortex (field Al) of the alert Mongolian gerbil. Eur J Neurosci 8:1001–1017
27. Ohl FW, Scheich H (1997) Learning-induced dynamic receptive field changes in primary auditory cortex of the unanaesthetized Mongolian gerbil. J Comp Physiol A 181:685–696
28. Ohl FW, Scheich H (2005) Learning-induced plasticity in animal and human auditory cortex. Curr Opin Neurobiol 15:470–477
29. Ohl FW, Wetzel W, Wagner T, Rech A, Scheich H (1999) Bilateral ablation of auditory cortex in Mongolian gerbil affects discrimination of frequency modulated tones but not of pure tones. Learn Mem 6:347–362
30. Ohl FW, Scheich H, Freeman WJ (2000) Topographic analysis of epidural pure-tone-evoked potentials in gerbil auditory cortex. J Neurophysiol 83:3123–3132
31. Ohl FW, Schulze H, Scheich H, Freeman WJ (2000) Spatial representation of frequency-modulated tones in gerbil auditory cortex revealed by epidural electrocorticography. J Physiol Paris 94:549–554
32. Ohl FW, Scheich H, Freeman WJ (2001) Change in pattern of ongoing cortical activity with auditory category learning. Nature 412:733–736
33. Ohl FW, Deliano M, Scheich H, Freeman WJ (2003) Early and late patterns of stimulus-related activity in auditory cortex of trained animals. Biol Cybern 88:374–379

34. Pribram KH (1991) Brain and perception: holonomy and structure in figural processing. Lawrence Erlbaum Associates, New Jersey
35. Pribram K (2007) Holonomic brain theory. Scholarpedia 2(5):2735
36. Scheich H, Brechmann A, Brosch M, Budinger E, Ohl FW (2007) The cognitive auditory cortex: task-specificity of stimulus representations. Hear Res 229:213–224
37. Schreiner CE, Polley DB (2014) Auditory map plasticity: diversity in causes and consequences. Curr Opin Neurobiol 24:143–156
38. Weinberger NM (2004) Specific long-term memory traces in primary auditory cortex. Nat Rev Neurosci 5:279–290
39. Weinberger NM, Hopkins W, Diamond DM (1984) Physiological plasticity of single neurons in auditory cortex of the cat during acquisition of the pupillary conditioned response: I. Primary field (AI). Behav Neurosci 98:171–188
40. Wetzel W, Wagner T, Ohl FW, Scheich H (1998) Categorical discrimination of direction in frequency-modulated tones by Mongolian gerbils. Behav Brain Res 91:29–39
41. Wetzel W1, Ohl FW, Scheich H (2008) Global versus local processing of frequency-modulated tones in gerbils: an animal model of lateralized auditory cortex functions. Proc Natl Acad Sci USA 105:6753–6758
42. Winer JA, Schreiner CE (eds) (2011) The auditory cortex. Springer, New York
43. Wright JJ, Bourke PD (2013) On the dynamics of cortical development: synchrony and synaptic self-organization. Front Comput Neurosci 7:4

Part IV
Commentaries on Differential Equation in Cortical Models

Chapter 15
Commentary by James J. Wright

Electrocortical Synchrony and the Regulation of Information Flow in the Cerebral Cortex

James J. Wright

Abstract A proposal is advanced, placing Freeman's work on cortical dynamics in relation to other work on brain organization. The cerebral cortex is treated as a stochastic, essentially linear, wave medium, the travelling waves generated by, and coordinating, transient foci of bursting, non-linear, gamma oscillation. During the bursts, information stored in synapses on distal dendritic trees of excitatory neurons is input to travelling waves. Interaction of the waves leads to transient steady-states, each approaching global synchrony, and coordinating pulse activity over the cortex. Each approach to global synchrony also determines sequential interactions of the cortex with subcortical systems, thus selecting sequential subsets of cortical state-space. It is further supposed that pre-synaptic competition for scarce resources within each neuron contributes to maximization of population synchrony. This entanglement of pulse/wave and synaptic dynamics permits a very large set of different transient organizations of pulse activity, and therefore cognitive states. These states can be represented within a closely related embryological model for cortical connectivity. Interaction of cortex and brain-stem also offers means of self-supervision of learning, via linkage to primitive motivational systems that can selectively regulate Hebbian synaptic consolidation.

15.1 Introduction

As of yet, agreed principles of brain function able to account for flexible, self-supervised, adaptive learning, elude us. What are the minimal biological properties that are essential to define the brain's basic operation? How does the brain store, access, and avoid mutual interference of information flows, while learning selectively?

Sir Charles Sherrington [22] recognized essential elements of these problems. He identified reflex systems, innately wired at lower brain levels, and operating with

---

J.J. Wright
Department of Psychological Medicine, University of Auckland
School of Medicine, Auckland, New Zealand

R. Kozma and W.J. Freeman, *Cognitive Phase Transitions in the Cerebral Cortex – Enhancing the Neuron Doctrine by Modeling Neural Fields*,
Studies in Systems, Decision and Control 39, DOI 10.1007/978-3-319-24406-8_15

reciprocal inhibition, so as to prevent confusion of separate responses. He thought similar principles might, during the evolutionary process, become extended into the cortex creating "a magic loom" [23].

Donald Hebb [11] envisaged transient functional "cell assemblies" of cortical neurons, blending into the next assembly in a "phase sequence". His learning rule for the development of synapses suggested how cell assemblies could be strengthened with practice, and how phase sequences are learned in order. Hebb did not specify the organization of cell assemblies or phase sequences in other than a general way. Howsoever the components are organized, newly evolved parts of the brain must be built upon, and interact with, the lower reflex processes, as was elucidated by MacLean [14]. Aspects of reptilian brains carry over into mammalian brains, as if the older brain was enveloped within the newer, with old and new acting in concert.

Walter Freeman's work [5–8] has characterized neuronal group properties, and revealed the occurrence of bursting activity in the gamma frequency range, associated with the perception of a stimulus, as opposed to direct sensory response. He and allied workers have identified widespread synchronous firing of neurons during the perceptual process and also the generation of expanding, contracting, and spiraling, cones of wave motion. He has characterized basic properties of the neural group response—a pulse-to-wave dendritic summation, and a wave-to-pulse asymmetrical sigmoid function governing the output response of neurons.

How then, might Sherrington's innate hard wired circuits, Hebb's flexible, evanescent and self-modifying networks, and MacLean's phylogenetically derived overall organization be understood in the light of Freeman's findings? Simulations of electrocortical activity go some way to solve this problem. They reproduce Freeman's findings, while introducing other essential biological components. Related recent theoretical work on cortical embryogenesis also suggests a reference framework upon which Freeman's cortical dynamics can operate. The elements of a proposal for their unification are diagramed in Fig. 15.1.

15.2 Neural Mean-Field Equations

Equations for the mean neural field, similar to those of Freeman, include those of [1, 16, 25, 26, 39] and the author's group [29, 30]. These can be considered as variants on a generic form of integral equations:

$$\varphi_p^{\mathbf{qr}}(t) = f_p^{\mathbf{qr}} \times Q_p\left(\mathbf{r}, \mathbf{t} - \frac{|\mathbf{q} - \mathbf{r}|}{\mathbf{v}}\right) \tag{15.1}$$

$$\psi_p^{\mathbf{qr}}(t) = M_p^{\mathbf{qp}} \times \varphi_p^{qp}(t) \tag{15.2}$$

$$\Psi_p(\mathbf{q}, t) = \int_r \psi_{\mathbf{p}}^{\mathbf{qr}}(t) d\mathbf{r} \tag{15.3}$$

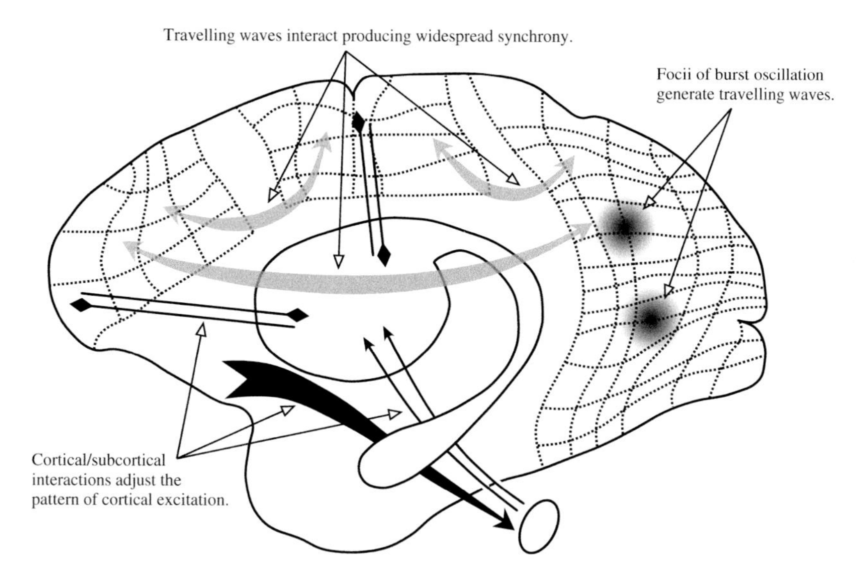

Fig. 15.1 Electrocortical activity—a punctuated equilibrium: Foci of excited neurons in cortex undergo non-linear bursting, during which retrograde propagation of action potentials into distal dendritic trees activates distal synapses, regulating the firing pattern. Surrounding the foci, near-linear travelling waves stimulate or suppress other foci. Selection of in-phase wave components, and synaptic competition for metabolic supply, approach a steady-state, with maximum synchrony at zero lag. Via cortical/subcortical interactions, the cortical steady state shifts the pattern of cortical excitation away from steady-state, renewing the cycle—and may also supervise synaptic consolidation. *Solid black arrows* represent interactions via axonal connections. *Shaded gray* pathways indicate polysynaptic, travelling wave, interactions (*White headed arrows* are merely label indicators)

$$V_p(\mathbf{q}, \mathbf{t}) = \mathbf{G_e} \times \Psi_\mathbf{e}(\mathbf{q}, \mathbf{t}) + \mathbf{G_i} \times \Psi_\mathbf{i}(\mathbf{q}, \mathbf{t}) \tag{15.4}$$

$$Q_p(\mathbf{q}, \mathbf{t}) = \mathbf{f}_\Sigma(\mathbf{V_p}(\mathbf{q}, \mathbf{t})) + \mathbf{E_p}(\mathbf{q}, \mathbf{t}) \tag{15.5}$$

Subscript $p = e, i$ refers to excitatory or inhibitory neurons; superscript **qr** refers to synaptic connection from **r** to **q** where **q**, **r** are cortical positions occupied by single neurons.

- $\varphi_p^{\mathbf{qr}}(t)$ is the flux of pulses reaching pre-synapses at the neuron at **q**, from the neuron at **r**.
- $\psi_p^{\mathbf{qr}}(t)$ is the synaptic current generated by $\varphi_p^{\mathbf{qr}}(t)$.
- $\Psi_p(\mathbf{q}, \mathbf{t})$ is the aggregate synaptic current of type p generated at **q**.
- $V_p(\mathbf{q}, \mathbf{t})$ is the soma membrane potential (relative to the resting potential) generated at **q**.
- $Q_p(\mathbf{q}, \mathbf{t})$ is the pulse emission rate at **q**.

- $f_p^{\mathbf{qr}}$ is the probability density of occurrence of pre-synapses generated by axons of the neuron at $\mathbf{r}$ terminating at $\mathbf{q}$.
- v is axonal conduction speed.
- $M_p^{\mathbf{qp}}$ is the steady-state term in a convolution transforming pre-synaptic flux to synaptic current.
- G_p is the steady-state term in a convolution transforming pre-synaptic flux into dendritic potentials.
- $f_\Sigma(V_p(\mathbf{q}, \mathbf{t}))$ is a sigmoid function describing the local conversion of dendritic potentials into the rate of generation of action potentials.
- $E_p(\mathbf{q}, \mathbf{t})$ is a driving signal noise, arising from intrinsic random cell action potentials, and applied inputs.

These equations represent the cortex in two dimensions as a mesh of excitatory and inhibitory neurons, with long-range excitation and short-range inhibition. With appropriate parameterization they enable reproduction of the $1/f^\alpha$ background EEG spectrum, and bursting oscillation in the beta/gamma range, with the mean value of $E_p(\mathbf{q}, \mathbf{t})$, representing the level of non-specific cortical excitation, regulating the transition between the $1/f$ background and the bursting oscillation. They sustain propagation of travelling waves, and synchronous oscillation [30].

With the introduction of lagged interaction with thalamic neurons they accurately reproduce the characteristic spectral sequence of the EEG's theta, alpha, beta and gamma background peaks [21] and permit simulation of evoked cortical responses [20].

At low pulse firing rates, corresponding to the resting cortex background state, these equations are linearized about small deviations low on the sigmoid wave-pulse function. At high firing rates, associated with bursting firing, they exceed threshold for undamped oscillation, and become significantly nonlinear.

15.3 Stochastic Equations in ODE Form

It may be objected that the mean-field may exert little variance on individual pulses. Although this may be so, there is still no reason for rejecting the physical validity of a linear treatment for low frequency activity in the cortex. The cortex can be described as composed of linked local excitatory/inhibitory interacting components:

$$\ddot{V}_e(\mathbf{q}, \mathbf{t}) + \mathbf{D}(\mathbf{q}, \mathbf{t})\dot{\mathbf{V}}_\mathbf{e}(\mathbf{q}, \mathbf{t}) + \mathbf{N}^2(\mathbf{q}, \mathbf{t})\mathbf{V}_\mathbf{e}(\mathbf{q}, \mathbf{t}) = \sum \mathbf{K}(\mathbf{q}, \mathbf{r}, \mathbf{t})\mathbf{V}_\mathbf{e}(\mathbf{r}, \mathbf{t}) \quad (15.6)$$

where $V_e(\mathbf{q}, \mathbf{t})$ are excitatory dendritic potentials (point local field potentials LFP), and $\{D, N, K\}$ are stochastic parameters. The complexities concealed in the constant parameter values of the field equations are implicit in the myriad and interdependent time-varying values of $\{D, N, K\}$.

Consequently, assuming extreme complexity in the factors governing parameters $\{D, N, K\}$, so that all are statistically independent, or have dependencies only within small groups, in sequential short epochs, it can be shown [28] that the natural frequencies of low frequency resonant modes have low time-variance, and their damping co-efficients are approximately equal at all frequencies, while high frequency resonant modes show high time-variance of both frequency and damping. Since the low frequency resonances approximate the time-courses of dendritic fields, and the high-frequency events approximate rapid events such as pulses, apparent dissociation of dendritic waves and individual pulses is expected.

Conversely, if sudden departures from steady-state occur, co-incident changes in members of $\{D, N, K\}$ will be reflected in highly nonlinear wave events.

This distinction permits the large-scale electrocortical field to be treated as a linear wave medium, driven by intrinsic episodes of burst firing, as well as by extrinsic specific, and non-specific inputs. The control and coordination of the global system can then be viewed as the interaction of the cortical and subcortical systems, and of local bursting interacting with the background field.

15.4 Cortical-Subcortical Interactions

Experimental evidence shows that the electrocortical field in a steady-state can be physically described as a near-equilibrium, multilinear system, regulated in part by cortical interactions with subcortical systems, as the stochastic equations suggest. Manipulation by lesion or stimulation of cortical activation pathways passing through the lateral hypothalamus—pathways exerting effects on alertness and motivation—reveals the electrocortical field to have constant natural frequencies, with the damping of the resonances increased by lateral hypothalamic input, along with an increase in the power of the noisy intrinsic signals driving the system [32, 33]. Autoregression analysis shows damping factors to be equal at all resonances for a given state of activation [35], and analysis of cortical evoked responses and their simulation indicate the evoked responses can be decomposed to superimposed impulse response functions [36].

Complex systems of pathways descending from cortex to the brain stem, passing through the hippocampus in part, and including connections to the reticular and non-specific thalamus may set cortical activation into one of many possible spatial patterns—these ongoing changes regulating attention and cognition. The discovery of hippocampal place neurons [17] may indicate the existence of specialised networks governing the stepping from one to another attentional set.

15.5 Pulse-Bursting and the Introduction of Stored Information

As shown by Freeman (*vide supra*), ongoing normal activity in cortex exhibits burst foci of 40 Hz activity, surrounded by travelling waves. Freeman associated the bursts with phase changes in the thermodynamic sense, and describes the travelling waves as phase-cones (phase in the angular sense) and vortices—the phase cones travelling either into a focus or away from a focus, often with multiple phase cones present at once.

Simulations of cortical activity [29, 30] can explain much of Freeman's findings. The foci of excitation are sites in which the excitatory/inhibitory local oscillation has become sufficiently excited to pass a threshold beyond which strong, non-linear oscillation is generated. Detailed cellular mechanisms remain to be fully specified within the simulations, but the following properties are apparent:

(i) The non-linear bursts can be self-stabilizing, as excitatory and inhibitory reversal potentials and synaptic adaptations can act to suppress the oscillation, thus limiting burst duration. Cortical-thalamic interactions could also act to stabilize the total field activity close to threshold for the onset/offset of oscillatory bursts.

(ii) Each focus transmits travelling waves into the essentially linear surrounding wave-medium, and depending on the frequency components of the inputs and the relative levels of excitation and inhibition, could either trigger or suppress other loci of oscillation. Simulations suggest that slow waves tend to suppress firing in surrounding areas, whereas activity near the resonant frequency would excite such activity. Higher frequency waves are more attenuated by spatial damping, so rich interactions between foci are possible.

(iii) During strong burst firing, retrograde propagation of action potentials has been shown to occur in the dendritic trees of the neurons [24]. This implies that distal synapses located on those dendritic trees, that were effectively inert, begin to play a role when the cells fire repeatedly during bursting. Conversely, wave transmission at low firing-rates must take place largely via synapses located on the proximal dendritic trees.

The consequence of these properties is that foci of oscillation and firing can communicate via travelling waves, and in so doing trigger and suppress activity at other sites, in complex patterns. In each local burst stored information is released into the electrocortical field from the otherwise quiescent distal-dendrite synapses—thus segregating synaptic information release into packages. Local and global self-stabilization mechanisms, held close to the threshold level for burst generation, thus sustain ongoing exchange of information cortex-wide.

15.6 Synchrony as the Global Attractor

A mechanism for the widespread occurrence of synchronous firing of pulses, and of the local field potential, is apparent in simulations, and arises as a universal property of networks with summing junctions [3]. It is generated by the constructive and destructive interference of travelling waves in this medium. That is to say—when pulse trains are summed in dendrites, in-phase components sum, and out-of-phase components are eliminated. Consequently, simulations show an equilibrium of signal exchange [31] in which excitatory neurons approach maximum zero lag synchrony develops rapidly. Excitatory and inhibitory cells approach zero-lag local correlation,[1] but with 1/4 cycle lag-correlation at greater distances of separation, conforming to results obtained by Freeman [5, 6]. As synchronous oscillation tends to a maximum, the free energy manifested in the travelling wave component tends to a minimum for that steady-state.

Thus, synchronous states act as global attractors for the electrocortical field, perturbations about the attractor can generate and coordinate the interaction of loci of pulse-bursts, and, by signaling subcortical systems, could direct changes in the cortical input conditions.

15.7 Stimulus-Feature-Linking, Phase Cones, Phase-Transitions, and Null-Spikes

Simulations [37] also reproduce properties of experimentally observed synchronous linking of cells responding to specific stimuli [11].

Travelling waves generated by any two bursting foci that have not yet come to equilibrium interact so that the less excited focus becomes the sink, and the more excited locus the source of waves, thus exhibiting the properties of "phase cones". More complex interactions could generate vortex-like wave appearances.

"Null-spikes"—that is values approaching zero in the Hilbert transform of a electrocortical voltage time-series—occur frequently and intermittently in electrocortical time-series, as reported by Freeman and colleagues. For a time-series $u(t)$, with time-step τ, the Hilbert transform is given by

$$H(u)(t) = \frac{1}{\pi} \lim_{\epsilon \to 0} \int_{\epsilon}^{\infty} \frac{u(t+\tau) + u(t-\tau)}{\tau} d\tau \tag{15.7}$$

Thus a null-spike is approached whenever signals at $t + \tau$ and $t - \tau$ approach mirror image, with reversal of sign. A step function, or bursts of oscillation can both

[1] For cells of each type that are closely situated, the same shared excitatory and inhibitory bursts are directly input to both cell types simultaneously, causing them to fire concurrently. At longer ranges, dendritic smoothing and the mixing of signals from multiple sources, allows the 1/4-cycle to-and-fro oscillation to occur in the large.

produce this effect, and may be concomitants of either local bursting, sudden-step "phase changes", or changes of local level of mean excitation generated by sudden changes of subcortical input—all with essentially stationary periods of electocortical activity either side of the transition, consistent with the occurrence of punctuated periods of equilibrium near a global attractor.

15.8 Information Capacity—Synapses and Their Developmental Organization

The wavelengths of electrocortical waves are very long, and their information capacity therefore limited but the cortex must be able to achieve a wide variety of output states. Coordinated modulation of synaptic gain, taking place on a time-scale slower than that of the electrocortical waves, may provide this second large-capacity informational channel.

Long, attenuated neurons, while providing large synaptic contact area of each neuron, place demand on transport systems within the cell to supply essential metabolites. There is good evidence that competition between synapses for resource takes place (e.g., [2, 10, 13, 15, 18] linking synaptic activity and resource uptake. If there is only enough of a crucial scarce metabolite for half the synapses to operate at maximum capacity, and if synapses must compete with each other on a small local scale, then, as well as generating a maximum of synapses, competition would maximize the possible Shannon entropy of the synaptic efficacies, thus, if the local supply of metabolites fluctuates with firing states of the network, and there are a multiplicity of essential scarce metabolic factors and their time-scales of supply, conditional Markov processes of great complexity, and a very large number of output pulse states, would be possible.

My colleagues and I have applied this idea to individual embryonic development of the cerebral cortex [34]. We proposed that rapidly dividing embryonic cortical neurons compete for limited metabolic resources and the unsuccessful competitors are eliminated by apoptosis (cell death). Survivors must emerge in arrays reducing the length of total axonal connections (thus reducing metabolic consumption), requiring selection of an array of neurons with ultra-small-world connectivity. We additionally proposed that synchronous firing of neurons serves to increase metabolic uptake per cell. From these assumptions, and with consolidation of the most long-term active synapses in accord with the Hebb rule, we have shown that, in the visual cortex, macrocolumns linked by superficial patchy connections must emerge in anatomically realistic patterns, with an antenatal arrangement which projects signals from the surrounding cortex onto each macrocolumn in a form analogous to the projection of a Euclidean plane onto a Möbius strip. This configuration reproduces typical cortical response maps. Simulations of signal flow explain cortical responses to moving lines as functions of stimulus velocity, length, and orientation not otherwise accounted for, and explains also certain seeming paradoxes of cortical organization before and after birth. The model can also be applied to the somato-sensory cortex [38] and the same principle appears likely to be applicable throughout all or most cortical areas. The assumptions of the model are supported by experimental findings that blocking synchronous firing increases cell apoptosis [12], while developing neurons in vitro form into small-world configurations, while firing synchronously [4].

There is no reason that the same relationship between synaptic resource competition and cell firing pertaining in the developing cortex should not apply also to the mature cortex—so this incidentally provides an account of the antenatal self-organization of cortical connections into a form ready to transmit organized images in any sensory mode, using the dynamics discussed earlier. Foci of bursting activity, releasing and receiving input information, and signals generated by sensory input and motor outputs could thus co-exist within a common reference frame.

15.9 Cortical Computation and Synchronous Fields

As activated cortical areas approach equilibrium with maximum synchrony, synaptic metabolic resources will become distributed to maximize the synchrony.

Substituting in Eq. (15.1)

$$f_p^{\mathbf{qr}} = \epsilon_p^{\mathbf{qr}} \times g_p^{\mathbf{qr}} \tag{15.8}$$

where

- $\epsilon_p^{\mathbf{qr}}$ is the pre-synaptic rate of metabolic supply, determining each synaptic efficacy, and follows the recent firing rate of a synapse connecting a neuron at **q** to a neuron at **r**. The total rate of supply is assumed to be sufficient to sustain only 50 % of a given neuron's synapses at maximum efficacy,
- $g_p^{\mathbf{qr}}$ is the synaptic density function for intra-cortical cells, composed of separate species of cells of differing characteristic axonal length, with summed range/synaptic density approximating ultra-small world connectivity,
- Within Eq. (15.5) subsets of cell pulse rates/densities arising from sensory and somatic inputs and nonspecific subcortical inputs can be introduced, along with intracortical groups of burst-firing cells and cortical cells firing slowly in their background state, as

$$\begin{aligned} Q_p(\mathbf{q}, t) = {} & Q(\text{bursting} - \text{cells}) + Q(\text{background}) \\ & + Q(\text{sensory} - \text{inputs}) + Q(\text{somatic} - \text{inputs}) \\ & + Q(\text{cortical/subcortical} - \text{return} - \text{inputs}) \end{aligned} \tag{15.9}$$

and introducing a definition for J, the joint synchrony of excitatory pre-synaptic pulses of mean rate $\bar{\varphi}_e$, throughout the field during a short epoch, T, when the mean pre-synaptic pulse rate is

$$J = \left[\frac{1}{T} \int_T \int_{\mathbf{r}} \int_{\mathbf{q}} (\varphi_e^{\mathbf{qr}} - \bar{\varphi}_e)^2 dt d\mathbf{r} d\mathbf{q} \right]^{1/2} \tag{15.10}$$

When a steady state is reached with J at maximum, half the synapses, with higher synaptic pulse densities will be receiving the scare resource at the optimum rate, and the other half of the synapses, with lower synaptic pulse densities, receiving effectively no resource. If there are N synapses, then there are $2^{N/2}$ ways the active and inactive synapses can be arranged—each a different configuration of cells firing with maximum zero-lag covariance, defining positions in an N-dimensional state-space of synaptic efficacies—all the possible steady-states of the cortex. Interference between these states, depending on their Mohalanobis distance, will segregate distant states from mutual interference, and tend to generalize states with small distance of separation, offering a mechanism of association.

The entangled synaptic metabolic dynamics and field pulse-and-wave dynamics would have the potential to perform large-scale parallel computations, in analogy to the way that quantum entanglement is envisaged to perform quantum computation—computations operating on sensory-motor sequences and associations, generating intentions and directing attention. Hebb's "cell assemblies" would correspond to pulse and synaptic efficacies as arrived at in each of the possible steady-states, and his "phase sequences" to the ongoing shifts of cortical activation mediated by the cortical-subcortical interactions. Assuming Hebb's rule—that synapses that are frequently active acquire increasing strength—permanent synaptic learning could emerge in the cortical computations. But without appropriate motivation and reinforcement these computations would remain disorganized and ineffective for organism survival.

15.10 Self-Supervision of Learning

The brain's inherited core structures mediating primitive survival behaviour emphasized by MacLean provide a means for self-supervision of learning. Powerful motivational effects, discovered by Olds and Milner [19], are exerted by stimulation of some subcortical structures—particularly pathways travelling to and from cortex via the lateral hypothalamus. Unilateral electrical stimulation of these motivational/reward pathways in "split-brain" animals, in which the great cerebral commissures were divided, has shown that the motivation systems concurrently consolidate learning in both of the divided cerebral hemispheres, despite separation of cognitive functions produced by the commissurotomy [27]. That is, the motivation pathways trigger storage of memory traces by a diffuse action, distinct from the localized storage of specific cognitive sequences in cortex itself.

It appears, therefore, that cognitive computations performed by entangled synaptic states and pulses, might be selectively stored on the basis of their cooperative consistency with hard-wired activity of lower-level survival circuits. If so, what the cortex learns is the pursuit of ancient patterns of survival behavior, to good effect, in ever increasingly complicated environments.

15.11 In Conclusion

This account goes some way to indicating how the brain might store, access, and avoid mutual interference of information flows, while learning selectively. The proposal is highly provisional, and requires extensive testing in competition with alternative models, wherever these are available. The proposed mechanisms of cortical and subcortical interaction leave undetermined how each steady-state is signaled from cortex to subcortex, and thus to the next subcortical input. The assumed link between synchronous cell firing and pre-synaptic metabolic competition requires more direct confirmation, and specification in biochemical detail, utilizing the large body of data already available. The proposed mechanism of self-supervision requires explication in terms of neuro-modulator function, and may be inadequate to explain all forms of learning.

That said, the proposal integrates otherwise independent lines of enquiry, and has the virtue of indicating means of natural selection for highly efficient connectivity, information storage, and massively parallel computation in the developing brain—both phylogenetically and ontologically.

References

1. Amari S (1971) Characteristics of randomly connected threshold-element networks and network systems. Proc IEEE 59:35–47
2. Barber MJ, Lichtman JW (1999) Activity-driven synapse elimination leads paradoxically to domination by inactive synapses. J Neurosci 19:9975–9985
3. Chapman CL, Bourke PD, Wright JJ (2002) Spatial eigenmodes and synchronous oscillation: coincidence detection in simulated cerebral cortex. J Math Biol 45:57–78
4. Downes JH, Hammond MW, Xydas D, Spencer M, Becerra VM, Warwick K, Whalley BJ, Nasuto SJ (2012) Emergence of a small-world functional network in cultured neurons. PLoS Comput Biol 8:e1002522. doi:10.1371/journal.pcbi.1002522
5. Freeman WJ (1975) Mass action in the nervous system. Academic Press, New York
6. Freeman WJ (1991) Predictions on neocortical dynamics derived from studies on paleocortex. Induced rhythms of the brain. Birkhauser, Boston
7. Freeman WJ, Barrie JM (2000) Analysis of spatial patterns of phase in neocortical gamma EEGs in rabbit. J Neurophysiol 84:1266–1278
8. Freeman WJ, Quiroga RQ (2013) Imaging brain function with EEG. Springer, New York
9. Gray CM, Konig P, Engel AK, Singer W (1989) Oscillatory responses in cat visual cortex exhibit intercolumnar synchronisation which reflects global stimulus properties. Nature 388:334–337
10. Hashimoto K, Tsujita M, Miyazaki T, Kitamura K, Yamazaki M, Shin H-S, Watanabe M, Sakimura K, Kano M (2011) Postsynaptic P/Q-type Ca2+ channel in Purkinji cell mediates synaptic competition and elimination in developing cerebellum. PNAS 108:9987–9992. doi:10.1073/pnas.1101488108

11. Hebb DO (1949) The organization of behavior. Wiley, New York
12. Heck N, Golbs A, Riedemann T, Sun J-J, Lessmann V, Luhmann HJ (2008) Activity dependent regulation of neuronal apoptosis in neonatal mouse cerebral cortex. Cereb Cortex 18:1335–1349
13. Kasthuri N, Lichtman JW (2003) The role of neuronal identity in synaptic competition. Nature 424:430. doi:10.1038/nature01836
14. MacLean PD (1973) A triune concept of the brain and behavior. In: Boag TJ, Campbell D (eds) The Hincks memorial lectures. University of Toronto Press, Toronto, pp 6–66
15. Miller KD (1996) Synaptic economics: competition and cooperation in correlation-based synaptic plasticity. Neuron 17:371–374
16. Nunez PL (1981) Electric fields of the brain: the neurophysics of EEG. Oxford University Press, New York
17. O'Keefe J, Dostrovsky J (1971) The Hippocampus as a spatial map. Preliminary evidence from unit activity in the freely-moving rat. Brain Res 34:171–175
18. Okamoto H, Ichikawa K (2000) A model for molecular mechanisms of synaptic competition for a finite resource. Biosystems 55:65–71
19. Olds J, Milner P (1954) Positive reinforcement produced by electrical stimulation of septal area and other regions of rat brain. Comp Physiol Psychol 47:419–427
20. Rennie CJ, Robinson PA, Wright JJ (2002) Unified neurophysical model of EEG spectra and evoked potentials. Biol Cybern 86:457–471
21. Robinson PA, Rennie CJ, Wright JJ, Bahramali H, Gordon E, Rowe DL (2001) Prediction of electroencephalographic spectra from neurophysiology. Phys Rev E 63(2):021903
22. Sherrington CS (1906) Activity-driven synapse elimination leads paradoxically to domination by inactive synapses. Yale University Press, New Haven
23. Sherrington CS (1940) Man on his nature. Gifford lecture. Cambridge University Press, Cambridge, pp 1937–1938
24. Stuart GJ, Sakmann B (1994) Active propagation of somatic action potentials into neocortical cell pyramidal dendrites. Nature 367:69–72
25. van Rotterdam A, Lopes da Silva FH, van den Ende J, Veirgever MA, Hermans AJ (1982) A model of the spatiotemporal characteristics of the alpha rhythm. Bull Math Biol 44:283–305
26. Wilson HR, Cowan J (1973) A mathematical theory of the functional dynamics of cortical and thalamic neurons tissue. Kybernetik 13:55–80
27. Wright JJ (1973) Unilateral pleasure-centre stimulation in split-brain cats. Exp Neurol 70:278–285
28. Wright JJ (1990) Reticular activation and the dynamics of neuronal networks. Biol Cybern 62:289–298
29. Wright JJ (2009) Cortical phase transitions: properties demonstrated in continuum simulations at mesoscopic and macroscopic scales. New Math Nat Comput 5:159–183
30. Wright JJ (2009) Generation and control of cortical gamma: findings from simulation at two scales. Neural Netw 22:373–384
31. Wright JJ (2011) Attractor dynamics and thermodynamic analogies in the cerebral cortex: synchronous oscillation, the background EEG, and the regulation of attention. Bull Math Biol. doi:10.1007/s11538-0109562-z
32. Wright JJ, Kydd RR (1984) A test for constant natural frequencies in electrocortical activity under lateral hypothalamic control. Biol Cybern 50:83–88
33. Wright JJ, Kydd RR (1984) Inference of a stable dispersion relation for electrocortical activity controlled by the lateral hypothalamus. Biol Cybern 50:88–94
34. Wright JJ, Bourke PD (2013) On the dynamics of cortical development: synchrony and synaptic self-organization. Front Comput Neurosci. doi:10.3389/fncom.2013.00004
35. Wright JJ, Kydd RR, Sergejew AA (1990) Autoregressive models of EEG. Results compared with expectation for a high-order multilinear near-equilibrium biophysical process. Biol Cybern 62:201–210
36. Wright JJ, Sergejew AA, Stampfer HJ (1990) Inverse filter computation of the neural impulse giving rise to the auditory evoked potential. Brain Topogr 2:293–302

37. Wright JJ, Bourke PD, Chapman CL (2000) Synchronous oscillation in the cerebral cortex and object coherence: simulation of basic electrophysiological findings. Biol Cybern 83:341–353
38. Wright JJ, Bourke PD, Favorov OV (2014) Möbius-strip-like columnar functional connections are revealed in somato-sensory receptive field centroids. Front Neuroanat. doi:10.3389/fnana.2014.00119
39. Zhadin MN (1982) Theory of rhythmic processes in the cerebral cortex. Academic Press, New York (in Russian)

Chapter 16
Commentary by Hans Liljenström

Multi-scale Causation in Brain Dynamics

Hans Liljenström

Abstract For any complex system, consisting of several organizational levels, the problem of causation is profound. Usually, science considers upward causation as fundamental, paying less or no attention to any downward causation. This is also true for the nervous system, where cortical neurodynamics and higher mental functions are normally considered causally dependent on the nerve cell activity, or even the activity at the ion channel level. This study presents both upward and downward causation in cortical neural systems, using computational methods with focus on cortical fluctuations. We have developed models of paleo- and neocortical structures, in order to study their mesoscopic neurodynamics, as a link between the microscopic neuronal and macroscopic mental events and processes. We demonstrate how both noise and chaos may play a role for the functions of cortical structures. While microscopic random noise may trigger meso- or macroscopic states, the nonlinear dynamics at these levels may also affect the activity at the microscopic level. We discuss some philosophical implications from these studies.

16.1 Introduction

The human brain is a complex system whose activity is reflected by a highly complex neurodynamics. This dynamics is characterized at a macroscopic level by oscillations, chaos and fluctuations, apparent in EEG and depending on underlying neural processes, external stimuli and various neuromodulatory mechanisms. The different organizational scales of the brain, from ion channels to neurons to net-works, are coupled via specific processes, each with a characteristic time scale.

Denis Noble has argued that there is no privileged level of causality in biological systems. Supposedly, "higher levels in biological systems exert their influence over the lower levels. Each level provides the boundary conditions under which the

H. Liljenström
Department of Energy and Technology, SLU, 7032, SE-75007 Uppsala, Sweden

H. Liljenström
Agora for Biosystems, 57, SE-19322 Sigtuna, Sweden

R. Kozma and W.J. Freeman, *Cognitive Phase Transitions in the Cerebral Cortex – Enhancing the Neuron Doctrine by Modeling Neural Fields*,
Studies in Systems, Decision and Control 39, DOI 10.1007/978-3-319-24406-8_16

processes at lower levels operate. Without boundary conditions, biological functions would not exist" [35]. The current work addresses this issue, with the aim of elucidating the causal pathways in brain dynamics, where downward causation from larger to smaller scales could be regarded as evidence that multi-level *both-way* causation occurs. While the outset for Noble's argument is the single (heart) cell with respect to its molecular constituents, the same arguments should apply to cortical networks and its cellular constituents.

For a nervous system, as for complex systems in general, different phenomena appear at different levels of aggregation. Emergent phenomena may result from a *bottom-up* causation, based on micro level phenomena. Yet, higher macro levels may also *control* lower ones (c.f. the so-called enslaving principle of [14]. This interplay between micro and macro levels is part of what frames the dynamics of neural systems. Of special interest is the meso level, i.e. the level in between the micro and the macro, where bottom-up meets top-down. The importance of the meso level in various sciences has been brought to attention in [32], where e.g. Hermann Haken emphasizes this level in physics, and Walter Freeman in neuroscience. Both Haken [14–17] and Freeman [4–6] have for long been dealing with the problem of *circular causality* in relation to micro, meso, and macro levels, but the problem still remains in discussions on how neural and mental processes are linked.

We use computational models of different brain structures, both of paleocortex and neocortex, to investigate how cortical neurodynamics may depend on structural properties, such as connectivity and neuronal types, and on intrinsic and external signals and fluctuations. In addition, we investigate to what extent the complex neurodynamics of cortical networks can influence the neural activity of single neurons. These issues have also been treated extensively by Walter Freeman and colleagues [5, 9].

Our results are suggestive for the neural mechanisms underlying EEG and the spatiotemporal patterns of activity associated with perception and cognitive functions, as well as for the dynamical effects of arousal and attention on cortical neurons. Our studies are also aiming at a greater understanding of the interplay be-tween order (e.g. in terms of regular oscillations) and disorder (noise and chaos) in neural information processing. In a larger context, this kind of studies should be relevant for our understanding of the intricate inter-relation between neural and mental processes, in particular the action-perception cycle [5, 22], which will be elaborated on in the Discussion section.

16.2 Cortical Network Models

For our studies, we use neural network models of paleocortex and neocortex, respectively, at different levels of details, depending on the particular questions and data available. These models are not connected in our modeling approach, but should rather be seen as examples of computational modeling applied to different problems and different cortical systems and processes. In some cases, we have been using models of the same system, but with different levels of details (with spiking neurons as network nodes, as well as population nodes), to investigate if there were any differences in the overall results. We could not find any qualitative differences in these cases.

16.2.1 Paleocortical Model

Our paleocortical model, which mimics the structures of hippocampus and the olfactory cortex, has network units with a continuous input-output function, corresponding to the average firing frequency of neural populations, which we compare with EEG and LFP data [24]. There are three cortical layers, with network units corresponding to populations of *feedforward* inhibitory interneurons, excitatory pyramidal cells, and feedback inhibitory interneurons, respectively. All connections are modeled with distance dependent time delays.

The time evolution for a network of N neural units is given by a set of coupled nonlinear first order differential delay equations for all the N internal states, u. With odor signal, $I(t)$, noise $\xi(t)$, characteristic time constant, τ_i, and connection weight w_{ij} between units i and j, separated by a time delay δ_{ij}, we have for each unit activity [24]:

$$\frac{du_i}{dt} = -\frac{u_i}{\tau_i} + \sum_{i \neq j}^{N} w_{ij} g_j [u_j(t - \delta_{ij})] + I_i(t) + \xi(t) \tag{16.1}$$

The input-output function, $g_j(u_j)$, is a continuous sigmoid function, experimentally determined by [2]:

$$g_i = AQ_i \left\{ 1 - exp\left[-\frac{exp(u_i) - 1}{Q_i} \right] \right\} \tag{16.2}$$

The gain parameter Q_i determines the slope, threshold and amplitude of the curve for unit i, and is associated with the level of arousal/attention as expressed through the level of acetylcholine. A is a normalization constant.

16.2.2 Neocortical Model

In our model of visual cortex, we use spiking model neurons, since we want to compare our results with observed data, as spike triggered averages of local field potentials. All model neurons satisfy the following Hodgkin-Huxley equation:

$$\begin{aligned} CV' = &-g_L(V + 67) - g_{Na} m^3 h(V - 50) - g_K n^4 (V + 100) \\ &- g_{AHP} w(V + 100) - I^{syn} + I^{appl} \end{aligned} \tag{16.3}$$

where V is the membrane potential and C is the membranecapacitance. g_L is the leak conductance, g_{Na} and g_K are the maximal sodium and potassium conductances, respectively. g_{AHP} is the maximal slow potassium conductance of the after hyperpolarization (AHP) current, which varies, depending on the attentional state. I^{syn} is the synaptic input current, and I^{appl} is the applied current. Variables m, h, n and w are calculated in a conventional way, and described more thoroughly in [13].

The six layers of visual cortex are lumped into three functional layers. In each of the three (lumped) layers of the local area network, there are four types of interactions: (1) lateral excitatory-excitatory, (2) excitatory-inhibitory, (3) inhibitory-excitatory, and (4) inhibitory-inhibitory, with corresponding connection strengths, which vary with distance between neurons (see [13] for details).

16.3 Simulation Results

16.3.1 Bottom-Up: Noise-Induced State Transitions

Noise appears primarily at the microscopic (subcellular and cellular) levels, but it is uncertain to what degree this noise normally is affecting meso- and macroscopic levels (networks and systems). Under certain circumstances, microscopic noise can induce effects on mesoscopic and macroscopic levels, but the role of these effects is still unclear. Evidence suggests that even single channel openings can cause intrinsic spontaneous impulse generation in a subset of small hippocampal neurons [20].

For a constant, low-amplitude random input (noise), the three-layered paleocortical network model is able to oscillate with two separate frequencies simultaneously, around 5 Hz (theta rhythm) and around 40 Hz (gamma rhythm). Under certain conditions, such as for high Q-values, the system can also display chaotic-like behavior, similar to that seen in EEG traces. In associative memory tasks, the network may initially display a chaotic-like dynamics, which then converges to a near limit cycle attractor, representing a stored memory (of an activity pattern) [25, 33].

Simulations with various noise levels show that spontaneously active neurons can induce global, synchronized oscillations with a frequency in the gamma range (30–70 Hz). Even if only a few network units are noisy, i.e. have an increased intrinsic random activity, and the rest are quiescent, coherent oscillatory or pseudo-chaotic activity can be induced in the entire network, if connection weights are large enough. The onset of global oscillatory/chaotic activity depends on, for example, connectivity, noise level, number of noisy units, and duration of the noise activity. The location and spatial distribution of these units in the network is also important for the onset and character of the global activity. For example, as the number or activity of the noisy units is increased, or if the distance between them increases, the oscillations tend to change into irregular patterns. In Fig. 16.1 we show that global network activity can be induced if only a small fraction of the network units are noisy (spontaneously active), and the rest are silent. After a short transient period of collective irregular

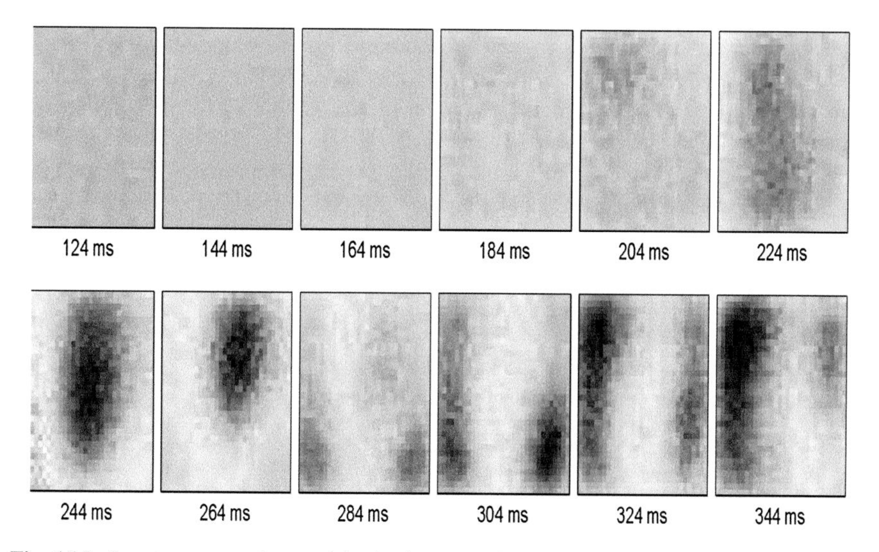

Fig. 16.1 Spontaneous random activity in the network may suddenly result in complex global activity patterns. The frames show snapshots 20 ms apart of the neural activity of the excitatory layer of the three-layered paleocortical network model, The model is here simulated with 32×32 network units in each of the three layers, corresponding to a 10×10 mm square of the real cortex

activity, the entire network begins to oscillate, and collective activity waves moves across the network. Similar effects can be obtained with regular oscillatory activity of a few network units (See ([28] and references therein).

16.3.2 Top-Down: Network Modulation of Neural Activity

The neural activity at the microscopic level of single neurons is the basis for the neurodynamics at the mesoscopic network level, and fluctuations may sometimes trigger coherent spatio-temporal patterns of activity at this higher level. Irregular chaotic-like behavior can be generated by the interplay of neural excitatory and inhibitory activity at the network level. This complex network dynamics, in turn, may influence the activity of single neurons, causing them to fire coherently or synchronously. This downward causation is complementary to the upward causation previously in focus.

Neromodulated Oscillations

The cortical neurodynamics observed in e.g. LFP and EEG studies may be (partly) controlled by neuromodulators, such as acetylcholine (ACh) and serotonin (5-HT). Such agents can change the excitability of a large number of neurons simultaneously, or the synaptic transmission between them. ACh is also known to increase the excitability by suppressing neuronal adaptation, an effect similar to that of increasing the gain in general. The concentration of these neuromodulators seems to be directly related to the arousal or motivation of the individual, and can have profound effects on the neural dynamics (e.g. an increased oscillatory activity) and on cognitive functions, such as associative memory [5, 31].

We use both our paleocortical and neocortical models, separately, for investigating how the network dynamics can be regulated by neuromodulators, implemented in the models as a varied excitability of the network units and modified connection strengths. The frequencies of the network oscillations depend primarily upon intrinsic time constants and delays, whereas the amplitudes depend predominantly upon connection weights and gains, which are under neuromodulatory control. Implementation of these neuromodulatory effects in both model systems cause dynamical changes analogous to those seen in physiological experiments.

Attention-Modulated Neurodynamics

Related to the level of arousal, and apparently also under neuromodulatory control, is the phenomenon of attention, which plays a key role in perception, action selection, object recognition and memory. The main effect of visual attentional selection appears to be a modulation of the underlying competitive interaction between stimuli in the visual field. Studies of cortical areas V2 and V4 indicate that attention modulates the suppressive interaction between two or more stimuli presented simultaneously within the receptive field [1]. Visual attention has several effects on modulating cortical oscillations, in terms of changes in firing rate [34], and gamma and beta coherence [11].

The inter-scale network interactions of various excitatory and inhibitory neurons in the visual cortex generate oscillatory signals with complex patterns of frequencies associated with particular states of the brain. Synchronous activity at an intermediate and lower-frequency range (theta, delta and alpha) between distant areas has been observed during perception of stimuli with varying behavioral significance [36, 38]. Rhythms in the beta (12–30 Hz) and the gamma (30–80 Hz) ranges are also found in the visual cortex, and are often associated with attention, perception, cognition and conscious awareness [12]. Data suggest that gamma rhythms are associated with relatively local connections, whereas beta rhythms are associated with higher level interactions.

Our simulation results show reduced beta synchronization with attention during a delay period (under certain modulation situations), and enhanced gamma synchronization, due to attention during a stimulation period (Fig. 16.2). In comparison with an idle state, where the dominant frequencies are around 17 Hz, the dominant frequency of the oscillatory synchronization and its STA (spike triggered averages) power in the *attended-in* group, A_{in}, is decreased, by inhibition of the intra-cortical synaptic inputs. This result agrees qualitatively with experimental findings that low-frequency synchronization is reduced during attention [38].

It is apparent that many factors play important roles in the network neurodynamics. These include (1) the interplay of ion channel dynamics and neuromodulation at a micro-scale, (2) lateral connection patterns within each layer, (3) feedforward and feedback connections between different layers at a meso-scale, and (4) top-down and

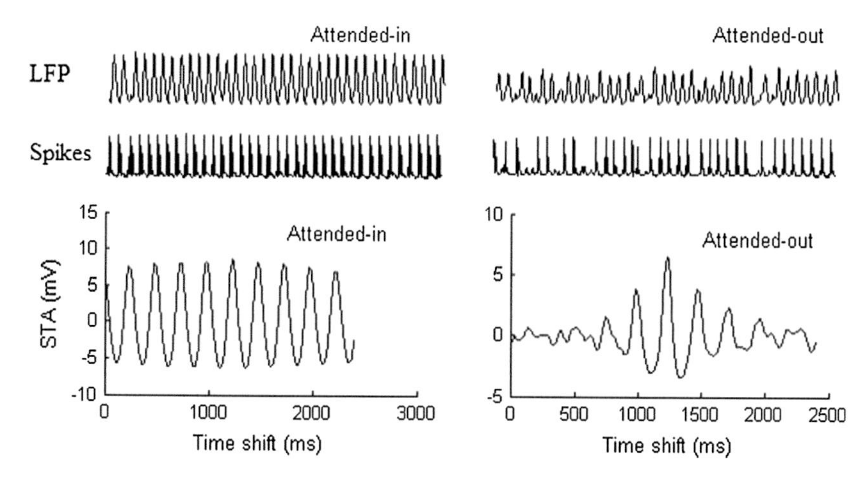

Fig. 16.2 Cholinergic modulation effects during a stimulus period. The activity of cortical neurons in attention (*left*) and out of attention (*right*). Local field potential (LFP), spikes, and spike triggered averages (STA) of attended-in and attended-out groups are shown, and calculated for the superficial layer of the six-layered neocortical network model; adopted from [13]

bottom-up circuitries at a macro-scale. The interaction between the top-down attention modulation, and the lateral short distance excitatory and long range inhibitory interactions, all contribute to the beta synchronization decrease during the delay period, and to the gamma synchronization enhancement during the stimulation period in the A_{in} group.

The top-down cholinergic modulation tends to enhance the excitability of the A_{in} group neurons. The Mexican hat shape lateral interactions mediate the competition between A_{in} and A_{out} (*attended-out*) groups. Other simulation results demonstrate (not shown) that the top-down attentional/cholinergic effects on individual neurons and on the local and global network connections are quite different. In particular, the higher beta synchronization of the A_{in} group is much stronger than that of the A_{out} group.

16.4 Discussion

Our simulations have demonstrated how events and processes at the microscopic level of single neurons can influence the mesoscopic neurodynamics of cortical networks, which in turn are associated with cognitive functions at the macroscopic level. It is apparent that internal noise can cause various phase transitions in the network dynamics, which may have effects on higher level functions. For example, an increased noise level in just a few network nodes can induce global synchronous oscillations in cortical networks and shift the system dynamics from one dynamical state to another. This in turn can change the efficiency in the information processing of the system. We have previously demonstrated that system performance can be maximized at an optimal noise level, analogous to the case of stochastic resonance, and that spontaneous activity can facilitate learning and associative memory [25, 26]. Thus, in addition to the (pseudo-) chaotic network dynamics, the noise produced by a few (or many) neurons, could make the system more flexible, increasing the responsiveness of the system and avoiding getting stuck in any undesired oscillatory mode [28, 29].

In addition, we have demonstrated how neuromodulation, whether related to the level of arousal or as a consequence of attention, can regulate the cortical neurodynamics, and hence the activity of its constituent neurons. The firing patterns of single neurons are thus, to a certain degree, determined by the activity at the net-work level (and above). For example, neurons in visual cortex may fire synchronously and in phase, as a result of cholinergic modulation during attention, which is confirmed by both experimental studies and computer simulations.

The objective has been to investigate how structure is related to dynamics, and how the dynamics at one scale is related to that of another. In this endeavor, we believe computational models of cortical structures can complement experimental studies in order to study the causal relationship between activities at different spatial and temporal scales.

Many of the issues we have been dealing with have also been treated by Freeman and colleagues [5, 10, 22, 23, 37], and there is much agreement between our results and interpretations, despite differences in model equations and network structures. Whereas Freeman and colleagues primarily base their modeling on the so-called K-sets [8], we have used leaky integrator equations of Hopfield type [18, 19] type for the neural populations of our paleocortical model, and Hodgkin-Huxley (and Fitzhugh-Nagumo) equations for spiking neurons of neocortex. The choice of models has been motivated by available experimental data and comprehensibility. A common feature of all these models is that they describe the oscillatory and/or chaotic dynamics of coupled excitatory and inhibitory neurons, apparent in LFP and EEG recordings.

All of these models attempt to capture some of the complexity of cortical net-works and their neurodynamics in order to elucidate how neural systems and processes could be linked to cognitive functions and consciousness, in particular to decision making and the action-perception cycle [3, 21, 27, 30], (See Fig. 16.3).

It is apparent that the intricate web of inter-relationships between different levels of neural organization, with inhibitory and excitatory feedforward and feed-back loops, with nonlinearities and thresholds, noise and chaos, makes any attempt to trace the causality of events and processes futile. In line with the ideas of [35], it seems obvious that there is, in general, both upward and downward causation in biological systems, including the nervous system. This also makes it impossible to say that mental processes are simply caused by neural processes, with-out any influence from the mental on the neural. On the contrary, these aspects of the human brain-mind relation seem complementary, and open up for a greater understanding of such ideas as *mind over matter*, placebo effects, and free will. Our minds may affect our bodies and environment, as much as vice versa.

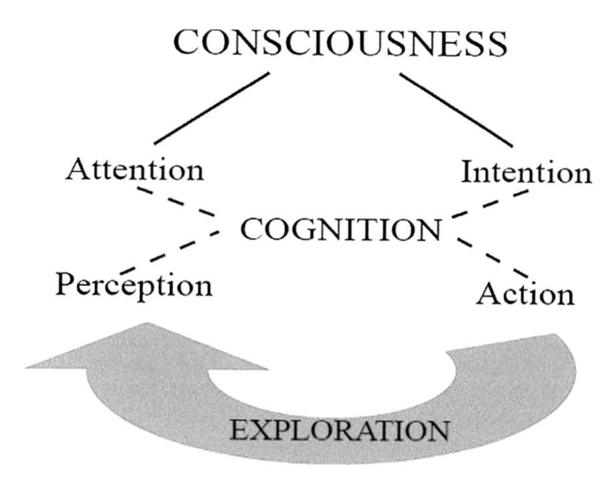

Fig. 16.3 Suggested relation between the dual aspects of consciousness (attention and intention), and the perception-action cycle in exploration of the world. There is no simple chain of causality in such a scheme, where each concept may be related to different interacting neural systems and processes; adopted from [30]

Acknowledgments I would like to thank my co-workers, Peter Århem, Yuqiao Gu, and Geir Halnes for fruitful collaboration. A grant (39987) from the John Templeton Foundation is gratefully acknowledged.

References

1. Corchs S, Deco G (2002) Large-scale neural model for visual attention: integration of experimental single-cell and fMRI data. Cereb Cortex 12:339–348
2. Freeman WJ (1979) Nonlinear gain mediating cortical stimulus-response relations. Biol Cybern 33:237–247
3. Freeman WJ (1995) Societies of brains - a study in the neurosciences of love and hate. Lawrence Erlbaum Assoc, Hillsdale
4. Freeman WJ (1999) Consciousness, intentionality, and causality. J Conscious Stud 6:143–172
5. Freeman WJ (2000) Neurodynamics: an exploration in mesoscopic brain dynamics. Springer, Berlin
6. Freeman WJ (2005) The necessity for mesoscopic organization to connect neural function to brain function. In: Liljenström H, Svedin U (eds) Micro - Meso - Macro: addressing complex systems couplings. World Scientific Publ. Co., Singapore, pp 25–36
7. Freeman WJ (2009) Nonlinear dynamics and intention according to Aquinas. Mind and Matter 6(2)
8. Freeman WJ, Erwin H (2008) Freeman K set. Scholarpedia 3(2):3238
9. Freeman WJ, Kozma R, Werbos PJ (2001) Biocomplexity: adaptive behavior in complex stochastic dynamical systems. Biosystems 59(2):109–123
10. Freeman WJ, Kozma R, Li G, Quiroga RQ, Vitiello G, Zhang T (2015) Advanced models of cortical dynamics in perception. In: Liljenström H (ed) Advances in cognitive neurodynamics, vol IV. Springer, Dordrecht, pp 127–136
11. Fries P, Reinolds JH, Rorje AE, Desimone R (2001) Modulation of oscillatory neuronal synchronization by selective visual attention. Science 291:1560–1563
12. Gray CM, König P, Engel A, Singer W (1989) Oscillatory responses in cat visual cortex exhibit inter-columnar synchronization which reflects global stimulus properties. Nature 338:334–337
13. Gu Y, Liljenström H (2007) A neural network model of attention-modulated neurodynamics. Cogn Neurodyn 1:275–285

14. Haken H (1983) Synergetics: an introduction. Springer, Berlin
15. Haken H (1991) Synergetic computers and cognition - a top-down approach to neural nets. Springer, Berlin
16. Haken H (1996) Principles of brain functioning. Springer, Berlin
17. Haken H (2005) Mesoscopic levels in science - some comments. In: Liljenström H, Svedin U (eds) Micro - Meso - Macro: addressing complex systems couplings. World Scientific Publ. Co., Singapore, pp 19–24
18. Hopfield JJ (1982) Neural networks and physical systems with emergent collective computational abilities. Proc Natl Acad Sci USA 79:2554–2558
19. Hopfield JJ (1984) Neurons with graded response have collective computational properties like those of two-state neurons. Proc Natl Acad Sci USA 81:3088–3092
20. Johansson S, Århem P (1994) Single-channel currents trigger action potentials in small cultured hippocampal neurons. Proc Natl Acad Sci USA 91:1761–1765
21. Kozma R, Freeman WJ (2009) The KIV model of intentional dynamics and decision making. Neural Netw 22(3):277–285
22. Kozma R, Freeman WJ, Érdi P (2003) The KIV model - nonlinear spatio-temporal dynamics of the primordial vertebrate forebrain. Neurocomputing 52:819–826
23. Kozma R, Puljic M, Balister P, Bollobás B, Freeman WJ (2005) Phase transitions in the neuropercolation model of neural populations with mixed local and non-local interactions. Biol Cybern 92(6):367–379
24. Liljenström H (1991) Modeling the dynamics of olfactory cortex using simplified network units and realistic architecture. Int J Neur Syst 2:1–15
25. Liljenström H (1995) Autonomous learning with complex dynamics. Intl J Intell Syst 10:119–153
26. Liljenström H (1996) Global effects of fluctuations in neural information processing. Int J Neur Syst 7:497–505
27. Liljenström H (1997) Cognition and the efficiency of neural processes. In: Århem P, Liljenström H, Svedin U (eds) Matter Maters - on the material basis of the cognitive activities of mind. Springer, Heidelberg
28. Liljenström H (2010a) Inducing phase transitions in mesoscopic brain dynamics. In: Steyn-Ross DA, Steyn-Ross ML (eds) Modeling phase transitions in the brain. Springer, New York, pp 147–175
29. Liljenström H (2010b) Network effects of synaptic modifications. Pharmacopsychiatry 43:S67–S81
30. Liljenström H (2011) Intention and attention in consciousness dynamics and evolution. J Cosmol 14:4848–4858
31. Liljenström H, Hasselmo ME (1995) Cholinergic modulation of cortical oscillatory dynamics. J Neurophysiol 74:288–297
32. Liljenström H, Svedin U (eds) (2005) Micro-Meso-Macro: addressing complex systems couplings. World Scientific, London
33. Liljenström H, Wu X (1995) Noise-enhanced performance in a cortical associative memory model. Int J Neur Syst 6:19–29
34. McAdams C, Maunsell J (1999) Effects of attention on orientation-tuning functions of single neurons in macaque cortical are V4. J Neurosci 19:431–441
35. Noble D (2012) A theory of biological relativity: no privileged level of causation. Interface Focus 2:55–64
36. Siegel M, Körding KP, König P (2000) Integrating top-down and bottom-up sensory processing by somato-dendritic interactions. J Comp Neurosci 8:161–173
37. Skarda CA, Freeman WJ (1987) How brains make chaos in order to make sense of the world. Brain Behav Sci 10:161–195
38. von Stein A, Chiang C, König P (2000) Top-down processing mediated by interareal synchronization. Proc Natl Acad Sci USA 97:14748–14753

Chapter 17
Commentary by Ray Brown and Morris Hirsch

Stretching and Folding in the KIII Neurodynamical Model

Ray Brown and Morris W. Hirsch

Abstract In this chapter we provide an alternate view of the KIII model derived from the laws of complexity (Stretching and Folding) rather than the laws of physics. This approach requires the use of Infinitesimal Diffeomorphisms (ID) in place of Ordinary Differential Equations. We indicate how IDs originate and then use them to replicate several examples from the work of Freeman and Kozma. By viewing the KIII theory as a purely mathematical system we anticipate that the KIII Theory will be made more accessible to researchers and scientists unfamiliar with the details of neuroscience and thus offer advances to the KIII Theory from other perspectives.

17.1 Introduction

Freeman and Kozma have introduced a paradigm shift in the analysis of neurodynamics by focusing on the mesoscopic structures external to the neurons referred to as the neuropil [1], rather than the dynamics of the neuronal mass only. An understanding of the amorphous nature of the neuropil, more analogous to a stiff fluid or a shag rug, suggested an entirely new approach to neurodynamical modeling that uses a field or wave paradigm as the means of communication, and the neuropil as the medium over which these waves must travel to relevant regions of the brain. It is on this fundamental wave-based neuropil approach that the KIII model is built. Importantly, their "wave" approach enables an explanation of how intentionality is communicated from the limbic system (the seat of intentionally) to specific regions of the brain in a manner that causes those regions to arrange or configure themselves to perform a desired new task. For example, the wave approach more efficiently

R. Brown
EEASI Corporation, 2100 Winrock Blvd. Suite 64, 77057 Houston, USA

M.W. Hirsch
University of Wisconsin at Madison and University of California at Berkeley, CA, USA

R. Kozma and W.J. Freeman, *Cognitive Phase Transitions in the Cerebral Cortex – Enhancing the Neuron Doctrine by Modeling Neural Fields*, Studies in Systems, Decision and Control 39, DOI 10.1007/978-3-319-24406-8_17

explains how one learns to hit a tennis ball for the first time given that they have never picked up a tennis racquet. This is because the range of dynamical activities involved in learning a new complex task must be performed in a few seconds and involves thought, action and emotion for the first time crossing the entire spectrum of human capability. Since mitosis is not operable within most of the neural mass (and certainly not rapidly), the problem of communicating intent and rapidly learning a new task driven by intent must have a dynamic that relies on speed. A wave dynamic is able to satisfy this specification.

In this volume, Vitiello describes two alternative approaches to advancing this theory based on the laws of physics. However, addressing how the dynamics, even at mesoscopic level, are transformed into everyday macroscopic behavior driven by human intentionality poses a significant problem which, as yet, has no solution. A fundamental road block is that the dynamics of everyday life cannot be formulated within the framework of the laws of physics. Thus we are stuck with the problem of bridging a dynamical system based on the laws of physics with a dynamical system most commonly described by statistics. In this chapter we introduce an entirely new approach to neuronal dynamics that side-steps the laws of Newton, physics and statistics. The approach presented here makes a more direct connection between the dynamics of the brain and the dynamics of humans at work and play by formulating both system within the same set of laws, the laws of complexity. The "laws" of complexity are found in the stretching and folding horseshoe paradigm of Smale [2]. This approach has been introduced in the analysis and simplification of physical systems by Hénon [3] in deriving his simplification of the dynamics of the Lorenz system (known now as the Hénon map); but a more extensive analysis is needed to apply this approach to both human and brain dynamics with equal legitimacy. An initial exploration of this concept is found in the Hirsch Conjecture [4] where it is noted that natural systems combine stretching and folding in very small increments as seen in ODEs having chaotic solutions. To obtain a general mathematical expression of complexity dynamics the concept of infinitesimal stretching and folding is introduced in [5]. To bridge the gap between neurodynamics and human dynamics at the macroscopic level it is noted that [5] the laws of complexity apply equally well to human dynamics and brain dynamics.

Vitielo also mentions another serious problem in understanding brain dynamics: The change in conductance of a single neuron cannot affect the dynamics of the mass action occurring inside the brain. To take this one step further, even minor changes at the mesoscopic level must not affect mass action dynamics. This aspect of neurodynamics is explained by two phenomena. (1) Any determination of the health of a human EEG is based on the morphological properties of the EEG, not its exact time series [6]. (2) The phenomena of sensitive dependence on initial conditions can only be understood in terms of the morphology of the time series in that two nearby trajectories may diverge or become uncorrelated, but their morphology does not change thus the mass action dynamics also do not change with small perturbations at the mesoscopic level [6].

In this chapter we shall refer to the KIII model as a bottoms-up approach. Using a bottoms-up approach, the laws of physics are applied to derive a system of differential equations known as the KIII model, Eq. 17.46. We will reverse this approach and use their work as a springboard to develop a top-down model that uses the “laws of complexity”, with which to derive their theory. It is hoped that the combined bottom-up physics approach and the top-down complexity approach will merge to produce an even greater methodology with which to analyze and perhaps prove theories of neurodynamics; and, through using the top-down approach, we hope to make research of the KIII theory available to a wide range of scientists and mathematicians who do not have an extensive background in neurodynamics, thermodynamics and quantum theory.

17.2 Stretching and Folding Provide an Alternative Approach to the Laws of Physics for Modeling Dynamics

The Newtonian approach to understanding and formulating equations of dynamics are expressed in his second law: $F = ma$. This formulation is excellent for physics but obscures the sources of complexity and chaos that can arise in dynamical systems generally. Hirsch in 1985 [7] set the stage for reexamination of the laws of Newton with this statement:

> A major challenge to mathematicians is to determine which dynamical systems are chaotic and which are not. Ideally one should be able to tell from the form of the differential equation (Morris W. Hirsch 1985 [7]).

In [5], following up on the conjecture of Hirsch, it is noted that by rearranging how the equations of dynamics are written (or simply viewing them from a different perspective), the form might be able to reveal the presence of chaos where the Newton approach of $F = ma$ does not. The key to doing this was the observation by Smale used in guiding the proof of the Smale Birkhoff Theorem [2] that the source of complexity arose by dividing the Newtonian forces into those that stretch and those that fold.

As an example, consider the Duffing/Ueda equation without damping:

$$\ddot{y} + y^3 = \beta \cos(t) \tag{17.1}$$

Written in Newtonian form we have

$$\ddot{y} = -y^3 + \beta \cos(t) \tag{17.2}$$

In this form, we are not led to sort out the source of complexity. Now we use parenthesis to group the terms as follows:

$$\ddot{y} = (-y^3 \text{ stretching}) + (\beta \cos(t) \text{ folding}) \tag{17.3}$$

Recognizing by definition of stretching and folding from [5] we see that $(\beta \cos(t))$ as the folding term and y^3 is the stretching term. We now know that the solution of this equation must be able to generate complexity not because $F = ma$, but because the forces involved are stretching and folding. If we apply this approach to the KIII model then we must arrange the differential equations into stretching and folding terms. The trick is to figure out how this translates into diffeomorphisms that contain all the complexity of the KIII model. The thermodynamical KIII model derives ODEs from the known dynamics of fluids moving across membranes; on the other hand, the ID model must begin with an identification of the stretching and folding components. From [4] we know that any diffeomorphism of the form

$$\mathbf{X} \to F(\mathbf{X}) \tag{17.4}$$

where $\nabla((\nabla \cdot F)(\mathbf{X})) \neq 0$ is a stretching component; and any diffeomorphism of the form

$$\mathbf{X} \to \exp(A) \cdot \mathbf{X} \tag{17.5}$$

where A is a $n \times n$ matrix of constants is a folding component.

There is one more step in the derivation. If we numerically integrate an ODE we must break down the numerical solution into discrete, but very short, steps. This implies that we must formulate stretching and fold in small steps, or "infinitesimal" increments. Using the concept of stretching and folding in small increments leads to the concept of Infinitesimal Diffeomorphisms as presented in [4]. By using infinitesimal steps we blend the dynamics of the two forces nearly continuously as often occurs in the natural world.

Now we must get an insight into how to transform an ODE into stretching and folding. There are two steps: (1) Recognizing the stretching and folding components in the ODE of interest; (2) Deriving how stretching and folding appear in ID by converting an ODE into an integral equation. To address (1) we use Eq. 3 from [1]:

$$\ddot{y_1} + \alpha\dot{y_1} + \beta y_1 = \beta w_{ee} Q(y_2) \tag{17.6}$$

$$\ddot{y_2} + \alpha\dot{y_2} + \beta y_2 = \beta w_{ee} Q(y_1) \tag{17.7}$$

Rearranging the equations into stretching and folding we have:

$$\ddot{y_1} = -(\alpha\dot{y_1} + \beta y_1 \text{ --folding}) + (\beta w_{ee} Q(y_2) \text{ --stretching}) \tag{17.8}$$

$$\ddot{y_2} = -(\alpha\dot{y_2} + \beta y_2 \text{ --folding}) + (\beta w_{ee} Q(y_1) \text{ --stretching}) \tag{17.9}$$

To address step (2), converting to an integral equation, we provide an intuitive derivation in the next section. Once the basic ideas are fixed, we may jump from ODEs to IDEs with a measure of ease. But first, we must emphasize another aspect of the ID formulation and its relationship to the KIII discrete dynamics.

The reformulation of the Freeman and Kozma KIII model as IDEs offers a serendipitous benefit that is related to the discrete dynamics of KIII. The IDE parameter h can vary from very small to quite large as shown in [5, 8]. This variation allow us to observe how the dynamics of the brain change with the degree of stretching and folding which, in turn, will be determined by external forces and intentionality. The Hénon study [8] and the studies in [5] show just how dramatic changes occur due to variations in the degree of stretching and folding. For example, in [8] the variation of the IDE parameter, h, can move the neurodynamics through a conventional period doubling process to chaos. One further aspect of the IDE formulation is that it allows us to the study how asymmetry in stretching and folding affects the neurodynamics. In this point-of-view there may be separate parameters, h_s, h_f, for stretching and folding that are driven by external and internal factors. The best example of this is found in [9] where the stretching dynamic is chosen to be a Bernoulli, or Anosov, map and the folding dynamic is an almost periodic map. In [9] they are combined as a weighted sum to demonstrate how remarkably the dynamics can vary as the weighting parameter is moved from 1 (only Bernoulli stretching) to 0 (almost periodic folding). Note that, in [9], the Bernoulli component can be further divided into a pure stretching and a pure folding component since Bernoulli is itself a consequence of stretching and folding.

17.3 Infinitesimal Diffeomorphisms First Originated from Integral Equations

Two theorems will serve to set the stage of the use of IDEs in biological systems generally. In [6] the ODE

$$\dot{x} + V(x) \cdot x = 0 \quad x(0) = x_0 \tag{17.10}$$

was introduced. An integral equation version of this ODE is given by

$$x(t) = \exp\left(-\int_{t_0}^{t} V(x(s))ds\right) \cdot x(t_0) \tag{17.11}$$

In higher dimensions $V(x)$ is a square matrix.

The importance of the form of this ODE is that it provides an entrance through which to understand the mathematical realization of stretching and folding. To better understand the ideas to come we will use a simplified version of Eq. 17.10:

$$\dot{x} + x = 0 \quad x(0) = x_0 \tag{17.12}$$

The solution is obviously $x(t) = \exp(-t) \cdot x_0$. Another way to view this solution is to set $t = h$, where h is a small step size.

$$x_{n+1} = \exp(-h) \cdot x_n \quad x_0 \text{ specified} \tag{17.13}$$

Now consider

$$\dot{x} + x = f(t) \quad x(0) = x_0 \tag{17.14}$$

This is converted into an integral equation by introducing an integrating factor $\exp(t)$

$$x(t) = \exp(-t)\,x_0 + \exp(-t)\int_0^t \exp(s) f(s) ds \tag{17.15}$$

$$x(t) = \exp(-t)\,x_0 + \exp(-t)\int_0^t f(s) d\exp(s) \tag{17.16}$$

Applying the mean value theorem to Eq. 17.16 we have may obtain a simple iteration scheme that will be presented in Theorem 17.2 to follow.

Let $x(t)$ be a real valued function of a real variable t. We have the following theorem concerning IDEs:

Theorem 17.1 *Assume*

$$\dot{x} + V(x)\cdot x = 0 \quad x(0) = x_0 \tag{17.17}$$

has a unique bounded solution for every initial condition and that $\|x(t)\| \leq M$ *for all* $t \in \mathbf{R}$. *Let* $t_n = n\,h$, *for* $h \in (0, a]$, *for* $a < 1$. *Also, define* $\tilde{x}$ *as*

$$\tilde{x}(t_{n+1}) = \exp(-V(x(t_n))(h))\cdot x(t_n) \tag{17.18}$$

then

$$\|x(t_n) - \tilde{x}(t_n)\| \leq K\cdot h \tag{17.19}$$

for fixed K *and all integers* n.

Proof A formal proof is deferred to [10]. We sketch some key steps to make the approach clear. The two primary steps are (1) to use mathematical induction to prove the approximation at the nth (the result is clearly true for $n = 0$); and, (2) to use mean value theorems to eliminate integrals in favor of algebraic terms.

Let $\Delta = \|x(t_n) - \tilde{x}(t_n)\|$ then the error at the nth step is given by

$$\Delta = \|(\exp(-V(x(\xi))\cdot h) - \exp(-V(x(t_0))\cdot h)))\cdot x(t_0)\| \tag{17.20}$$

and so

$$\Delta \leq \|\exp(-V(x(\xi))\cdot h) - \exp(-V(x(t_0))\cdot h)\|\cdot M \tag{17.21}$$

and

$$\Delta \leq \|\exp(-V(x(\rho)))\cdot h)\|\cdot\|V(x(t_0)) - V(x(\xi))\|\cdot h)\cdot M \tag{17.22}$$

Let

$$K_1 = \max_x \|V(x(t))\| \text{ and } K_2 = \max_x \|V'(x(t))\|$$

to get

$$\Delta \leq \|\exp(K_1\cdot h)\|\cdot K_2\|x(t_0) - x(\xi)\|\cdot h\cdot M \tag{17.23}$$

$$\Delta \leq \|\exp(K_1 \cdot h)\| \cdot K_2 \|V(x(t)) \cdot x(t) \cdot h\| \cdot h \cdot M \tag{17.24}$$

$$\Delta \leq \|\exp(K_1 \cdot h)\| \cdot K_2 \cdot K_1 \cdot h^2 \cdot M^2 \tag{17.25}$$

showing that the error can be made arbitrarily small. End of sketch.

In particular we have the iteration

$$x_{n+1} = \exp(-V(x_n) \cdot h)) \cdot x_n \tag{17.26}$$

as an approximation to the solution of Eq. 17.10 when all smoothness and boundedness assumptions are satisfied.

Now consider Eq. 17.27 with the same assumptions as Theorem 17.1. We roughly sketch the derivation of the relevant IDE and state the theorem afterwards.

$$\dot{x} + V(x) \cdot x = f(t) \quad x(0) = x_0 \tag{17.27}$$

By taking $f(t)$ to be a constant, b, over a very small interval $[t_n, t_{n+1}]$ we may obtain a integral equation containing a convolution. Note that the assumption on f implies that the derivative of f is not too troublesome, i.e., $\|f'(t)\|$ is uniformly bounded over the entire real line.

We need the substitution

$$\frac{d\,W(t)}{dt} = V(x(t))$$

Introducing an integrating factor into Eq. 17.27 and collecting terms we have

$$\frac{d(x(t)\exp(W(t))}{dt} = f(t)\exp(W(t)) \tag{17.28}$$

Integrating over a small interval $[t_n, t_{n+1}]$

$$x(t_{n+1})\exp(W(t_{n+1})) = x(t_n)\exp(W(t_n)) + \int_{t_n}^{t_{n+1}} f(s)\exp(W(s))ds \tag{17.29}$$

$$x(t_{n+1}) = x(t_n) \cdot \exp(-V(x_n)\,h) + \exp(-W(t)) \int_{t_n}^{t_{n+1}} f(s)\exp(W(s))ds \tag{17.30}$$

Since we have the solution for the homogeneous equation we only need to consider approximating the inhomogeneous part

$$\exp(-W(t_n)) \int_{t_n}^{t_{n+1}} f(s)\exp(W(s))ds \approx \exp(-W(t_n))\, b \int_{t_n}^{t_{n+1}} \exp(W(s))ds \tag{17.31}$$

where we have substituted b for $f(t)$ over the interval $[t_n, t_{n+1}]$. At this point we assume $W(t) = t$ to shorten and simplify the discussion so that it better applies to the following presentation. Then the integral can be explicitly evaluated and we arrive at the form of the IDE that will be present in the following discussion.

$$\exp(-(t_n))\, b \int_{t_n}^{t_{n+1}} \exp(s)ds = \exp(-(t_n))\, b\, (\exp(t_{n+1}) - \exp(t_n)) \tag{17.32}$$

This gives us $b\,(\exp(h) - 1)$ for Eq. 17.32. Collecting terms we have the theorem for the case where $V(x) = \alpha$.

Theorem 17.2 *The IDE for Eq. 17.27 is given by*

$$x_{n+1} = \exp(\alpha \cdot h)(x_n - f(n\,h)) + f(n\,h)$$

For $V(x)$ not constant, the derivation is more involved and can be found in [10].

17.4 Deriving IDEs for the KIII Model

Note that all equations of the KIII model may be represented in the general form:

$$\frac{d\mathbf{X}}{dt} = \mathbf{AX} + F(\mathbf{X}, t) \tag{17.33}$$

The origin of IDs comes from converting Eq. 17.33 to an integral equation and then simplifying. An intuitive derivation goes as follows:

$$\begin{aligned}\exp(-A \cdot (t+h))X(t+h) &= \exp(-A \cdot t)X(t) \\ &\quad + \int_t^{t+h} \exp(-As)F(X(s))ds \qquad (17.34)\\ X(t+h) &= \exp(A \cdot h)X(t) + \exp(A \cdot (t+h)) \\ &\quad \int_t^{t+h} \exp(-As)F(X(s))ds \\ &= \exp(A \cdot h)X(t) + \exp(A \cdot (t+h)) \\ &\quad \cdot \int_t^{t+h} (-A^{-1})F(X(s))d\exp(-As) \\ &= \exp(A \cdot h)X(t) + \exp(A \cdot (t+h))((-A^{-1})F(X(\xi)) \\ &\quad \cdot (\exp(-A(t+h) - \exp(At)) \\ &= \exp(A \cdot h)X(t) + (-A^{-1})F(X(\xi)(1 - \exp(A \cdot h))\end{aligned} \tag{17.35}$$

$$X(t+h) = \exp(A \cdot h)(X(t) + (A^{-1})F(X(\xi))(-A^{-1})F(X(\xi)) \tag{17.36}$$

$$X(t_{n+1}) \approx \exp(A \cdot h)(X(t_n) - G(X(t_n)) + G(X(t_n)) \tag{17.37}$$

where $G(X(t)) = (-A^{-1})F(X(\xi))$ and $exp = \exp(A \cdot h)$. This requires that A^{-1} exist. When the solution is an attractor, and F is bounded, the ID provides a very good approximation to the solution of a nonlinear autonomous ODE.

Using this form of the ID justifies looking for solutions to any equation of the form Eq. 17.33 by assuming it has the form of an IDE. The correspondence is this:

$$\dot{\mathbf{X}} = \mathbf{AX} + F(\mathbf{X}, t) \tag{17.38}$$

$$\mathbf{T}_h(\mathbf{X}) = \exp(\mathbf{B} \cdot h)(\mathbf{X} - G(\mathbf{X}, f(h))) + G(\mathbf{X}, f(h)) \tag{17.39}$$

$$\mathbf{X}_{n+1} = \exp(\mathbf{B} \cdot h)(\mathbf{X}_n - G(\mathbf{X}_n, n \cdot h))) + G(\mathbf{X}_n, n \cdot h) \tag{17.40}$$

This derivation is partly formal, partly experimental. In general, we start with the form of an ID if we are working with an equation of the form of Eq. 17.33 and then we use formal data to obtain the best approximation to the stretching terms and folding terms separately. The folding terms will be captured in the eigenvalues of **B** and the stretching terms will be determined by the form of G and its "stretching" parameters. The model derived by shifting our emphasis from KIII ODEs to KIII ID will be referred to as KIII-ID.

From an engineering point of view, since we are starting with the known form of the solution, using the KIII-ID the numerical approximation and modeling should be achieved with a significant reduction in computational effort. This may come in the form of a reduction in the number of equations needed to model neurodynamics. We justify abandoning the derivation of a specific time series related to the physics described by the ODEs by the morphology principle of the EEG. This recognizes that it is the "form" of the equations that will best capture neurodynamics rather than an analysis of the physics of fluids or quantum theory.

17.4.1 The Linear ID Provides Fundamental Insights into the Dynamics of Stretching and Folding Systems

A linear Infinitesimal Diffeomorphism (ID) is

$$\mathbf{T}_h(\mathbf{X}) = \exp(\mathbf{A} \cdot h)(\mathbf{X} - F(\mathbf{X})) + F(\mathbf{X}) \tag{17.41}$$

where $\mathbf{A} \neq 0$ is an n by n matrix of constants, $\mathbf{X}$ is an n-vector and $\nabla(\nabla \bullet F) \neq 0$, where F is twice differentiable function on $\mathbf{R}^n$ and $h \neq 0$.

The condition $\nabla(\nabla \bullet F) \neq 0$ is the definition of stretching.

A linear ID inherently combines stretching and folding infinitesimally through the step size h. The folding part is given by $\exp(\mathbf{A} \cdot h)(\mathbf{X})$ since $\nabla(\nabla \bullet (\exp(\mathbf{A} \cdot h)(\mathbf{X})) = 0$ and the stretching part is suppled by F by the condition $\nabla(\nabla \bullet F) \neq 0$.

Consider

$$\dot{\mathbf{X}} = \mathbf{AX} + F(\mathbf{X}) \tag{17.42}$$

If F is bounded, then Eq. 17.41 accurately approximates the solution of Eq. 17.42:

Theorem 17.3 *Let F if bounded on* $\mathbf{R}^n$ *then for* h_1, h_2, *then*

$$\|\mathbf{T}_{h_1}(\mathbf{X}) - \mathbf{T}_{h_2}(\mathbf{X})\| \leq K\|h_1 - h_2\|\|\mathbf{X}\| \tag{17.43}$$

for some constant K which depends on the bound of F.

The fixed points of $\mathbf{T}$ are given by

$$\mathbf{T}_h(\mathbf{X}) = \mathbf{X} \tag{17.44}$$

or

$$\exp(\mathbf{A} \cdot h)(\mathbf{X} - F(\mathbf{X})) = \mathbf{X} - F(\mathbf{X}) \tag{17.45}$$

Equation 17.45 implies $\mathbf{X} - F(\mathbf{X})$ belongs to the kernel of $\mathbf{A}$. Thus set of fixed points of the one-parameter family $\mathbf{T}_h$ is precisely the kernel of $\mathbf{A}$. Some of the fixed points of the linear ID are given by $F(\mathbf{X}) = \mathbf{X}$. The dynamics of the fixed points are given by the Jacobian of $\mathbf{T}$.

17.4.2 The Standard KIII Model Can Be Reformulated as a Set of Infinitesimal Diffeomorphisms (ID)

The standard thermodynamic KIII model can be described by a vector equation whose most general form is Eq. 17.46. Note that in [1] second order ODEs are used as a basis for formulating the KIII model. To translate this into IDs, we replace each second order ODE with a pair of IDs.

$$\frac{d\mathbf{X}}{dt} = \mathbf{AX} + F(\mathbf{X}, t) \tag{17.46}$$

The function $F(\mathbf{X}, t)$ is given as follows, see Eq. 8 of [1]:

$$F(\mathbf{X}, t) = \sum_{j}^{N} w_j Q(y_j) + \sum_{j}^{N} \sum_{\tau}^{T} k_{ijk} Q(y_j(t - \tau)) + P_i(t) \tag{17.47}$$

Equation 17.46 can be solved by iterating the vector mapping

$$\mathbf{X}_{n+1} = \exp(\mathbf{A} \cdot h)(\mathbf{X}_n - F(\mathbf{X}_n, t_n)) + F(\mathbf{X}_n, t_n) \tag{17.48}$$

and when the matrix **A** has some eigenvalues less than 1, this approximation can be extremely accurate, see [12], and the step size h can be as large as 0.5 while retaining the morphological properties of the exact solution, [4]. Note that in Eq. 17.48 the time variable may be absorbed to make the equation *autonomous* by increasing the dimension by 1.

Equation 17.48 is required to have sufficiently smooth derivatives. We rewrite Eq. 17.48 in the form of a transformation on a manifold:

$$\mathbf{T}_h(\mathbf{X}) = \exp(\mathbf{A} \cdot h)(\mathbf{X}_n - F(\mathbf{X}_n)) + F(\mathbf{X}_n) \tag{17.49}$$

The ID, Eq. 17.49, has broad applicability and occurs in a wide range of problems of physics, fluid flow and electronic circuits [13].

More generally, an Infinitesimal Diffeomorphism (ID) is a one-parameter family of maps on $\mathbf{R}^n$ of the form (17.50) where F is a twice differentiable mapping from $\mathbf{R}^n$ to $\mathbf{R}^n$, $G(X)$ is a twice differentiable matrix function of $X \in \mathbf{R}^n$ and $h \neq 0$ is a real parameter.

$$\mathbf{T}_h(\mathbf{X}) = \exp(G(\mathbf{X}) \cdot h)(\mathbf{X} - F(\mathbf{X})) + F(\mathbf{X}) \tag{17.50}$$

As noted previously, the significance of IDs is that they are diffeomorphisms that also have the characteristics of a time series. This fact makes it possible to analyze very complex nonlinear processes more efficiently than by using conventional numerical methods. In addition to the ability to analyze fundamental dynamics, the ID provides an avenue for compression of high-dimensional systems of ODEs due to its similarity to Gaussian integration for second-order ODEs. IDs are particularly well suited to analyze the morphology of nonlinear ODEs of the form (17.46) which includes such equations as the Chua double scroll, the Lorenz system, the Rössler system and the K-neurodynamical models that will be discussed in this paper.

17.5 The Application of IDs to K-Neurodynamics May Result in Useful Simplifications of the ODEs Use to Describe the KIII System

In this section will apply IDs to formulate the K-neurodynamical models. These models will be designated as the K-ID models.

The K0-ID infinitesimal diffeomorphism is a direct translation of the K0 Eq. 1 model [1]. As noted earlier, this translations replaces a single second order ODE with a pair of IDs.

$$\mathbf{X}_n = ((x_n - F(x_n, y_n)); \ \mathbf{Y}_n = ((y_n - F(x_n, y_n)) \tag{17.51}$$

$$\begin{pmatrix} x_{n+1} \\ y_{n+1} \end{pmatrix} = \begin{pmatrix} \exp(\alpha \cdot h) \cdot \mathbf{X}_n \cdot \cos(\omega \cdot h) + \mathbf{Y}_n \cdot \sin(\omega \cdot h)) + F(x_n, y_n) \\ \exp(\alpha \cdot h) \cdot \mathbf{Y}_n \cdot \cos(\omega \cdot h) - \mathbf{X}_n \cdot \sin(\omega \cdot h)) + F(x_n, y_n) \end{pmatrix} \tag{17.52}$$

The KI-ID infinitesimal diffeomorphism is a modified version of the KI model, Eq. 3 model in [1] Some abbreviations are needed here:

$$\mathbf{X}_n = ((x_n - F(z_n, w_n)); \ \mathbf{Y}_n = ((y_n - F(z_n, w_n)) \tag{17.53}$$

$$\mathbf{Z}_n = ((z_n - F(x_n, y_n)); \ \mathbf{W}_n = ((w_n - F(x_n, y_n)) \tag{17.54}$$

$$\begin{pmatrix} x_{n+1} \\ y_{n+1} \\ z_{n+1} \\ w_{n+1} \end{pmatrix} = \begin{pmatrix} \exp(\alpha \cdot h) \cdot \mathbf{X}_n \cdot \cos(\omega \cdot h) + \mathbf{Y}_n \cdot \sin(\omega \cdot h)) + F(z_n, w_n) \\ \exp(\alpha \cdot h) \cdot \mathbf{Y}_n \cdot \cos(\omega \cdot h) - \mathbf{X}_n \cdot \sin(\omega \cdot h)) + F(z_n, w_n) \\ \exp(\alpha \cdot h) \cdot \mathbf{Z}_n \cdot \cos(\omega \cdot h) + \mathbf{W}_n \cdot \sin(\omega \cdot h)) + F(x_n, y_n) \\ \exp(\alpha \cdot h) \cdot \mathbf{W}_n \cdot \cos(\omega \cdot h) - \mathbf{Z}_n \cdot \sin(\omega \cdot h)) + F(x_n, y_n) \end{pmatrix} \tag{17.55}$$

Rewriting the above equations in the terminology of [1] we have, with the following abbreviations

$$\mathbf{X}_n = ((x_n - Q(v)); \ \mathbf{Y}_n = ((y_n - Q(v)) \tag{17.56}$$

$$\mathbf{Z}_n = ((z_n - Q(u)); \ \mathbf{W}_n = ((w_n - Q(u)) \tag{17.57}$$

$$\begin{pmatrix} x_{n+1} \\ y_{n+1} \\ z_{n+1} \\ w_{n+1} \end{pmatrix} = \begin{pmatrix} \exp(\alpha \cdot h) \cdot \mathbf{X}_n \cdot \cos(\omega \cdot h) + \mathbf{Y}_n \cdot \sin(\omega \cdot h)) + Q(v) \\ \exp(\alpha \cdot h) \cdot \mathbf{Y}_n \cdot \cos(\omega \cdot h) - \mathbf{X}_n \cdot \sin(\omega \cdot h)) + Q(v) \\ \exp(\alpha \cdot h) \cdot \mathbf{Z}_n \cdot \cos(\omega \cdot h) + \mathbf{W}_n \cdot \sin(\omega \cdot h)) + Q(u) \\ \exp(\alpha \cdot h) \cdot \mathbf{W}_n \cdot \cos(\omega \cdot h) - \mathbf{Z}_n \cdot \sin(\omega \cdot h)) + Q(u) \end{pmatrix} \tag{17.58}$$

where

$$Q(s) = q_m \cdot \left(1 - \exp\left(\frac{1 - \exp(s)}{q_m}\right)\right) \tag{17.59}$$

and $v = x - 5.23 \cdot w \cdot z$ and $u = y - 0.1 \cdot x$ and $q_m = 5.0$

Again, we see that two second order ODEs are replaced by 4 IDs.

17.6 The KIII-ID Model Can Provide a Reduction in Computation as Well as Insights into the Neurodynamics

We now define the KIII-ID as follows. Assume that Eq. 17.60 is true.

$$\sum_{j}^{N}\sum_{\tau}^{T} k_{ijk}Q(y_j(t-\tau)) \approx Q(f(y_1, y_2, \dots y_N)) = \Phi(\mathbf{X}) \tag{17.60}$$

and let $\mathcal{Q}$ be defined as follows:

$$\mathcal{Q} = \sum_{j}^{N} w_j Q(y_j) \tag{17.61}$$

Let $\Psi(\mathbf{X}) = \mathcal{Q}(\mathbf{X}) + Q(f(y_1, y_2, \dots y_N) = \mathcal{Q}(\mathbf{X}) + \Phi(\mathbf{X})$. Then KIII-ID is given by

$$\mathbf{T}_h(\mathbf{X}) = \exp(\mathbf{A} \cdot h)(\mathbf{X} - \Psi(\mathbf{X})) + \Psi(\mathbf{X}) \tag{17.62}$$

The mesoscopic theory requires a wave to pulse dynamic to communicate intent to local regions of the brain responsible for initiating action quickly. This wave dynamic may be what is referred to as a calcium wave in [12] that moves through the neuropil. This leads us to conjecture that the KIII-ID model can be further abstracted by the introduction of a wave/field concept. To arrive at the KIII-ID field model we make the following assumptions:

1. The KIII model was derived from the Neurodynamics of the brain using the simplest possible approach that captures the essential features of EEG studies.
2. The actual dynamics of the brain are so complex that it is reasonable to try to abstract from the KIII model only the essential concepts and dynamics inherent in that model assuming a field theory of the brain.
3. Then derive an abstract model of KIII by reverse engineering the KIII ODE model. To do this, two modification to the KIII theory were introduced: (A) in place of ODEs we used IDs that provide a dramatic simplification of the Runge-Kutta integration approach while retaining all dynamics and providing for step size variation without any loss of morphological accuracy. (B) Assume that the forcing function of KIII model, which was derived by direct experimentation, must be morphologically equivalent to a field force having a much simpler form.
4. $\Psi(f(\mathbf{X}))$ is the field-composite of all interactions between nodes of the KII model. In terms of ID theory, f will represent the transition surface in n-dimensional space which governs the stretching wave in the neuropil. $\exp(\mathbf{A} \cdot h)$ will provide the folding wave component.

Given these abstractions, we present a simulation of the KII-ID model and the KI-ID model and contrast their morphology with EEG recordings from [14]. The

Table 17.1 Data for the KI-ID system in Fig. 17.1a

Damping and frequency	$\alpha = -0.1 : \beta = 0.5$
Step size	$h = 0.001$
Number of iterations	$\mathrm{N} = 1,000,000$
Initial conditions	$x = 0 : y = 1 : z = 0 : w = 1.5$

Table 17.2 Parameters of the KII-ID model in Fig. 17.1b

Damping and frequency	$\alpha = -0.08 : \beta = 0.9$
Step size	$h = 0.001$
Number of iterations	$\mathrm{N} = 1,000,000$
Initial conditions	$x = 0 : y = 1 : z = 0 : w = 1.5$
Initial conditions	$x_1 = 0 : y_1 = 1 : z_1 = 1 : w_1 = 1.5$

Table 17.3 Parameters for KIII-ID shown in Fig. 17.2

Damping and frequency	$\alpha = -0.03 : \beta = 0.5$
Step size	$h = 0.01$
Number of iterations	$\mathrm{N} = 1,000,000$
Initial conditions	$x = 0 : y = 0.02 : z = 0 : w = 0.05$
Initial conditions	$x_1 = 0 : y_1 = 0.2 : z_1 = 0 : w_1 = 0.5$
Initial conditions	$x_2 = 0 : y_2 = 0.2 : z_2 = 0 : w_2 = 0.5$

simulations are an abstraction of Eqs. (3) and (13) from [1] (Tables 17.1, 17.2 and 17.3).

Iteration equations for KI-ID are as follows:

$$Q_0 = Q(x - 5.23 \cdot w \cdot z) \tag{17.63}$$

$$Q_1 = Q(y - 0.1 \cdot x) \tag{17.64}$$

$$Q(v) = 5.0 \cdot (1 - \exp((1 - \exp(v))/5.0))) \tag{17.65}$$

$$x \to \exp(\alpha \cdot h) \cdot ((x - Q_1) \cdot \cos(\beta \cdot h) + (y - Q_1) \cdot \sin(\beta \cdot h)) + Q_1 \tag{17.66}$$

$$y \to \exp(\alpha \cdot h) \cdot ((y - Q_1) \cdot \cos(\beta \cdot h) - (x - Q_1) \cdot \sin(\beta \cdot h)) + Q_1 \tag{17.67}$$

$$z \to \exp(\alpha \cdot h) \cdot ((z - Q_0) \cdot \cos(\beta \cdot h) + (w - Q_0) \cdot \sin(\beta \cdot h)) + Q_0 \tag{17.68}$$

$$w \to \exp(\alpha \cdot h) \cdot ((w - Q_0) \cdot \cos(\beta \cdot h) - (z - Q_0) \cdot \sin(\beta \cdot h)) + Q_0 \tag{17.69}$$

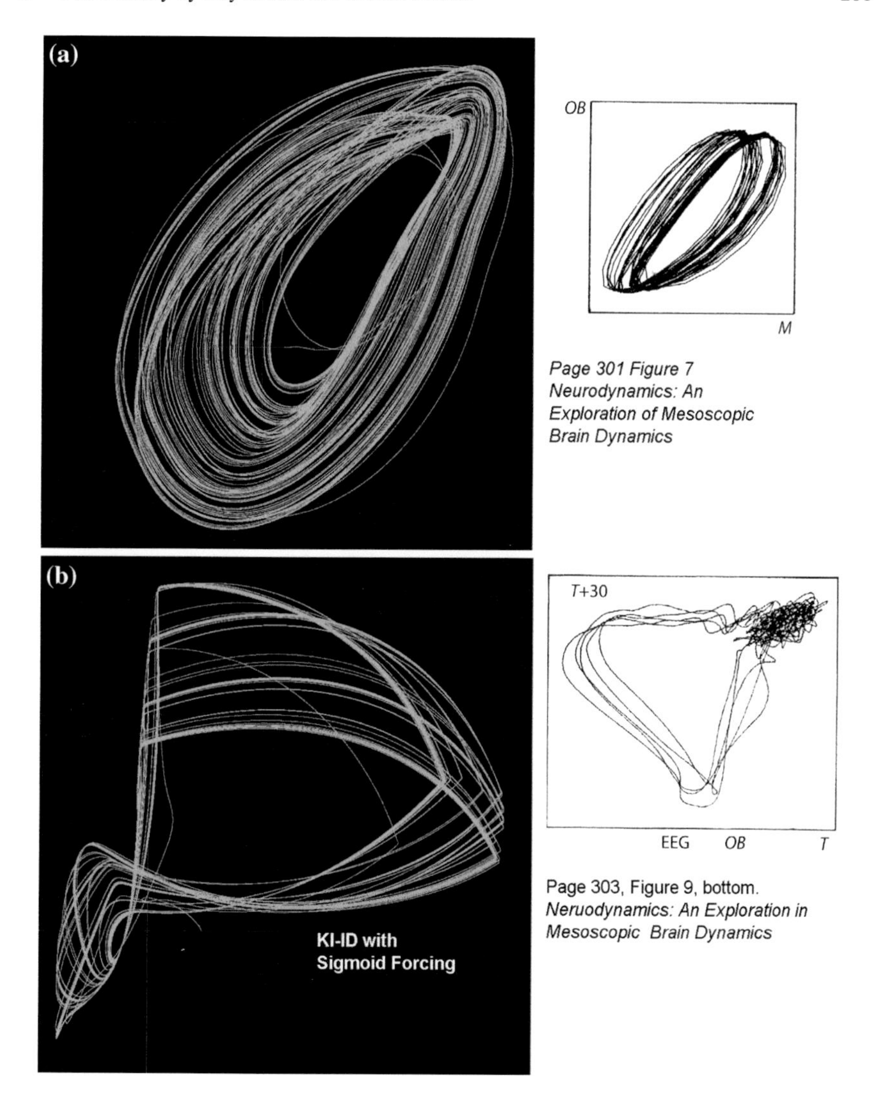

Fig. 17.1 Plate A Page 301, Fig. 7, *Left Plate* from Freeman [14], Plate B (p. 303), Fig. 9, *bottom*

Iteration equations for KII-ID are as follows:

$$Q_0 = Q(y_1) + 0.6 \cdot Q(z) \tag{17.70}$$

$$Q_1 = Q(y_1 + w_1) - Q(z) \tag{17.71}$$

$$Q_3 = Q(y - w) + 1.1 \cdot Q(z) \tag{17.72}$$

$$Q_4 = Q(y - x) \tag{17.73}$$

$$Q(v) = 5.0 \cdot (1 - \exp((1 - \exp(v))/5.0))) \tag{17.74}$$

$$x \rightarrow \exp(\alpha \cdot h) \cdot ((x - Q_1) \cdot \cos(\beta \cdot h) + (y - Q_1) \cdot \sin(\beta \cdot h)) + Q_1 \tag{17.75}$$

$$y \rightarrow \exp(\alpha \cdot h) \cdot ((y - Q_1) \cdot \cos(\beta \cdot h) - (x - Q_1) \cdot \sin(\beta \cdot h)) + Q_1 \tag{17.76}$$

$$z \to \exp(\alpha \cdot h) \cdot ((z - Q_0) \cdot \cos(\beta \cdot h) + (w - Q_0) \cdot \sin(\beta \cdot h)) + Q_0 \quad (17.77)$$
$$w \to \exp(\alpha \cdot h) \cdot ((w - Q_0) \cdot \cos(\beta \cdot h) - (z - Q_0) \cdot \sin(\beta \cdot h)) + Q_0 \quad (17.78)$$
$$x_1 \to \exp(\alpha \cdot h) \cdot ((x_1 - Q_3) \cdot \cos(\beta \cdot h) + (y_1 - Q_3) \cdot \sin(\beta \cdot h)) + Q_3 \quad (17.79)$$
$$y_1 \to \exp(\alpha \cdot h) \cdot ((y_1 - Q_3) \cdot \cos(\beta \cdot h) - (x_1 - Q_3) \cdot \sin(\beta \cdot h)) + Q_3 \quad (17.80)$$
$$z_1 \to \exp(\alpha \cdot h) \cdot ((z_1 - Q_4) \cdot \cos(\beta \cdot h) + (w_1 - Q_4) \cdot \sin(\beta \cdot h)) + Q_4 \quad (17.81)$$
$$w_1 \to \exp(\alpha \cdot h) \cdot ((w_1 - Q_4) \cdot \cos(\beta \cdot h) - (z_1 - Q_4) \cdot \sin(\beta \cdot h)) + Q_4 \quad (17.82)$$

17.7 The Wave $\Psi(X)$ for Any K Model May Arise from Partial Differential Equations that Must Be Derived from Experiment

While the sigmoid function is known to describe neuron binary dynamics, the complex summation of sigmoid functions could be replaced by a morphologically equivalent function which is known to satisfy a wave equation, for example $\sin(u) + \sin(3 \cdot u)/3 + \cdots$. Figure 17.2 compares using a wave-sigmoid dynamic to a Global wave dynamic in the KII-ID model.

In the KIII-ID model, the function $Q(v)$ which represents the transfer from a wave to an impulse is replaced by a new function that collectively describes the local dynamics without considering the specific dynamics of wave-pulse interaction. This is a mathematical abstraction and simplification to break from the physics and a transition to just the collective dynamics of all forces and interactions combined. Making this abstraction alleviates the researcher unskilled in neuroscience from fully understanding the particulars of the wave-to-pulse dynamic and only considering mathematical dynamics. While this does place the engineer a step away from the neuroscience, it may also facilitate formulations that will encompass additional insights and provide access to the KIII theory by scientists and engineers unskilled in the details of neuroscience.

Iteration equations for KIII-ID are as follows:

$$Q_0 = Q(w) + Q(x_2 + \Psi(x_2, w_1, z, w)) \quad (17.83)$$
$$Q_1 = Q(y_1 + w_1) - Q(z) \quad (17.84)$$
$$Q_3 = Q(y - w) + 1.1 \cdot Q(z) \quad (17.85)$$
$$Q_4 = Q(y - x) \quad (17.86)$$
$$Q(v) = (1 - \exp((1 - \exp(v)))) \quad \text{Plate A} \quad (17.87)$$
$$Q(v) = \sin(v) + \sin(3 \cdot v)/3 \quad \text{Plate B} \quad (17.88)$$
$$\Psi(x, y, z, w) = \exp(\alpha \cdot x) \cdot \cos(\alpha \cdot y) + \sin(5 \cdot z) \cdot \cos(\cos(x) \cdot w) \quad (17.89)$$
$$x \to \exp(\alpha \cdot h) \cdot ((x - Q_1) \cdot \cos(\beta \cdot h) \quad (17.90)$$
$$+ (y - Q_1) \cdot \sin(\beta \cdot h)) + Q_1 \quad (17.91)$$
$$y \to \exp(\alpha \cdot h) \cdot ((y - Q_1) \cdot \cos(\beta \cdot h) \quad (17.92)$$

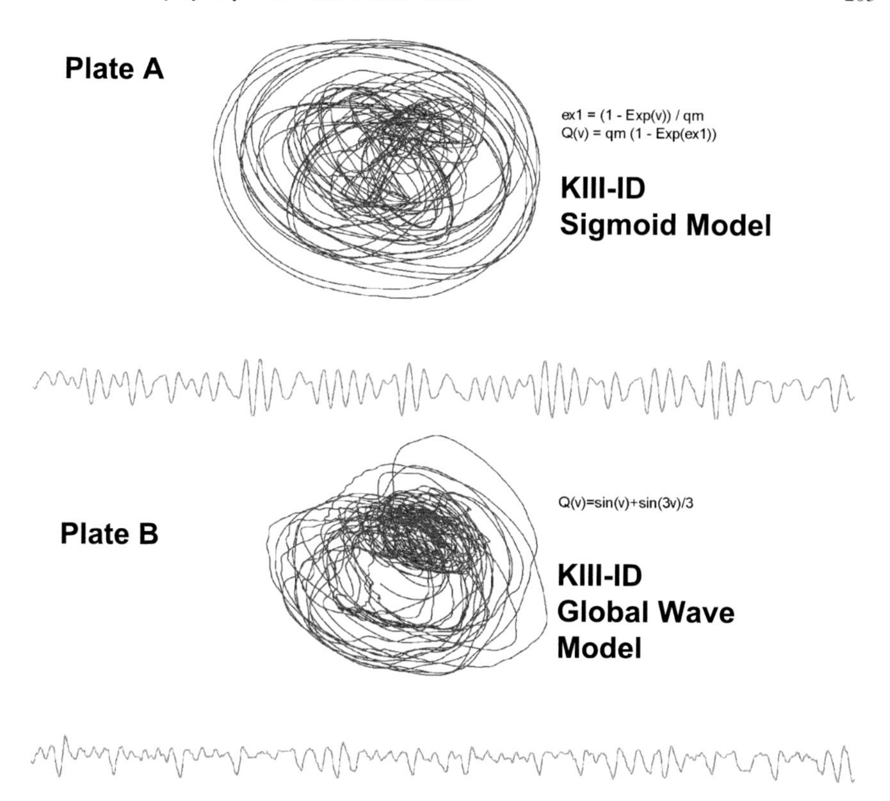

Fig. 17.2 KIII-ID Sigmoid model versus the KIII-ID wave model

$$- (x - Q_1) \cdot \sin(\beta \cdot h)) + Q_1 \tag{17.93}$$
$$z \to \exp(\alpha \cdot h) \cdot ((z - Q_0) \cdot \cos(\beta \cdot h) \tag{17.94}$$
$$+ (w - Q_0) \cdot \sin(\beta \cdot h)) + Q_0 \tag{17.95}$$
$$w \to \exp(\alpha \cdot h) \cdot ((w - Q_0) \cdot \cos(\beta \cdot h) \tag{17.96}$$
$$- (z - Q_0) \cdot \sin(\beta \cdot h)) + Q_0 \tag{17.97}$$
$$x_1 \to \exp(\alpha \cdot h) \cdot ((x_1 - Q_3) \cdot \cos(\beta \cdot h) \tag{17.98}$$
$$+ (y_1 - Q_3) \cdot \sin(\beta \cdot h)) + Q_3 \tag{17.99}$$
$$y_1 \to \exp(\alpha \cdot h) \cdot ((y_1 - Q_3) \cdot \cos(\beta \cdot h) \tag{17.100}$$
$$- (x_1 - Q_3) \cdot \sin(\beta \cdot h)) + Q_3 \tag{17.101}$$
$$z_1 \to \exp(\alpha \cdot h) \cdot ((z_1 - Q_2) \cdot \cos(\beta \cdot h) \tag{17.102}$$
$$+ (w_1 - Q_2) \cdot \sin(\beta \cdot h)) + Q_2 \tag{17.103}$$
$$w_1 \to \exp(\alpha \cdot h) \cdot ((w_1 - Q_2) \cdot \cos(\beta \cdot h) \tag{17.104}$$
$$- (z_1 - Q_2) \cdot \sin(\beta \cdot h)) + Q_2 \tag{17.105}$$
$$x_2 \to \exp(\alpha \cdot h) \cdot ((x_2 - Q_5) \cdot \cos(\beta \cdot h) \tag{17.106}$$

$$+ (y_2 - Q_5) \cdot \sin(\beta \cdot h)) + Q_5 \tag{17.107}$$

$$y_2 \rightarrow \exp(\alpha \cdot h) \cdot ((y_2 - Q_5) \cdot \cos(\beta \cdot h) \tag{17.108}$$

$$- (x_2 - Q_5) \cdot \sin(\beta \cdot h)) + Q_5 \tag{17.109}$$

$$z_2 \rightarrow \exp(\alpha \cdot h) \cdot ((z_2 - Q_4) \cdot \cos(\beta \cdot h) \tag{17.110}$$

$$+ (w_2 - Q_4) \cdot \sin(\beta \cdot h)) + Q_4 \tag{17.111}$$

$$w_2 \rightarrow \exp(\alpha \cdot h) \cdot ((w_2 - Q_4) \cdot \cos(\beta \cdot h) \tag{17.112}$$

$$- (z_2 - Q_4) \cdot \sin(\beta \cdot h)) + Q_4 \tag{17.113}$$

17.8 Summary

Starting with the KIII wave theory of Freeman-Kozma, we derived a top-down mathematical model, KIII-ID, which used the concept of stretching and folding in place of the laws of physics. We noted that the ID model has a mathematical foundation that has broad applicability to many dynamical systems including the KIII ODEs. We discussed some of the simplifying advantages of the KIII-ID approach and then we used the KII-ID model to morphologically replicate results from the KIII model of known EEGs. Finally we suggested that the sigmoid function could be replaced by solutions of wave equations which may lead to further simplifications of the KIII theory and make it more accessible to researchers without an extensive knowledge of neurodynamics as well as more amenable to formal scientific proof.

References

1. Ilin R, Kozma R (2006) Stability of coupled excitatory-inhibitory neural populations and applications to control of multi-stable systems. Phys Lett A 360:66–83
2. Smale S (1967) Differentiable dynamical systems. Bull Am Math Soc 73:747–817
3. Hénon M (1976) A two-dimensional mapping with a strange attractor. Commun Math Phys 50(1):69–77
4. Brown R (2014) The Hirsch conjecture. Dynamics of continuous, discrete and impulsive systems (to appear)
5. Brown R (2015) The theory of infinitesimal diffeomorphisms: an introduction (to appear)
6. Brown R, Jain V (2009) A new approach to Chaos. Dyn Contin Discret Impuls Syst Ser A Math Anal 16:863–890
7. Hirsch MW (1985) Fischer P, Smith W (eds) The chaos of dynamical systems, chaos, fractals and dynamics. Marcel Decker, Inc, New York
8. Brown R (2015) Stretching and Folding Transitions in the Infinitesimal Hénon Diffeomorphism (to appear)
9. Brown R, Chua L (1996) From almost periodic to chaotic: the fundamental map. Int J Bifurc Chaos 6(6):1111–1125
10. Brown R (2015) A short history of stretching and folding (to appear)
11. Brown R (2014) Infinitesimal stretching and folding II: the simple scroll. Dyn Contin Discret Impuls Syst (to appear)
12. Brown R (2014) Infinitesimal stretching and folding I (to appear)
13. Brown R (1992) Generalizations of the Chua equations. Int J Bifurc Chaos 4:889–909
14. Freeman WJ (2000) Neurodynamics: an exploration in mesoscopic brain dynamics. Springer, New York

Chapter 18
Commentary by Ray Brown on Real World Applications

An Essay on the Use of Sports Training (Tennis) to Prove Experimentally a Theory of Brain Dynamics

Ray Brown

Abstract In this work, we examine the basic problem of connecting a theory of the brain to the activities of humans engaged in the common pursuits of everyday life. This examination is explored through an implementation of a current dynamical theory, the KIII theory, which originates with Freeman [3, 4] and is advanced mathematically by Kozma. Our venue is a sports training program which is chosen for its accessibility to all researchers. In order to carry out this examination we must use a mathematical framework that serves the purpose of capturing the dynamics of the Freeman-Kozma model (Freeman, Neurodynamics: an exploration in mesoscopic brain dynamics, 2000, [4], Ilin and Kozma, Phys Lett A 360:66–83, 2006, [6]) and which can also be applied to the activities of a human enterprise.

18.1 Introduction

> *And men ought to know that from nothing else but thence [from the brain] come joys, delights, laughter and sports, and sorrows, griefs, despondency, and lamentations. And by this, in an especial manner, we acquire wisdom and knowledge, and see and hear, and know what are foul and what are fair, what are bad and what are good, what are sweet, and what unsavory... And by the same organ we become mad and delirious, and fears and terrors assail us... in these ways I am of the opinion that the brain exercises the greatest power in the man.*—Hippocrates [5]

What is at issue today in the quote of Hippocrates is this: If we succeed in constructing a dynamical model of the brain, how do we connect these dynamics to the behavior and actions of humans engaged in normal, everyday activities as Hippocrates claims? That is the central focus of this article.

R. Brown
EEASI Corporation, 2100 Winrock Blvd. Suite 64, Houston, TX 77057, USA

R. Kozma and W.J. Freeman, *Cognitive Phase Transitions in the Cerebral Cortex – Enhancing the Neuron Doctrine by Modeling Neural Fields*, Studies in Systems, Decision and Control 39, DOI 10.1007/978-3-319-24406-8_18

The sum total of our knowledge of the human brain arises from (1) experiments on animals; (2) medical cases of humans with some form of brain trauma; (3) statistical correlations from studies of psychology and sociology. However, direct quantifiable examination and study of normal humans engaged in normal activities is what would best confirm a theory. Specifically, direct quantifiable cause and effect data, rather than statistical correlations, are what is needed for sufficient proof of a neurodynamical theory.

18.2 Implementation of the KIII Model

To "prove" a neurodynamical theory it would be sufficient, in addition to accounting for laboratory experimental results, to answer the three key questions in Table 18.1.

Any program to prove a neurodynamical theory must be able to use normal human subjects performing normal human activities. To establish such a program three conditions must be met: (1) Find a simplified venue within which the key issues that bar our ability to answer these questions can be examined; (2) apply our knowledge of the dynamics of the human brain theory to this venue to construct a minimal system to test and prove hypotheses that will provide the foundation on which further developments can be based; (3) establish specifications for the design of systems which are broad enough to encompass a wide range of human enterprises.

In this paper I will select a venue for testing the KIII Theory and construct, using that venue, a minimal system that will address the three conditions cited above.

18.3 Selection of Mesoscopic Components

We begin by tabulating the minimal factors that must be instantiated in the sports training program based on an understanding of the KIII Theory and its implications:

While most of the factors in Table 18.2 are self explanatory, the factors of Fear/stress and ARTT require some explanation. A wave-pulse mesoscopic theory allows for rapid response to surprise, fear, stress and circumstances that challenge the limits of human performance. To test the KIII Theory it was essential to challenge the limits of human performance and endurance, particularly in responding to fast-paced events. To do this we introduced many stress related activities that contributed directly to student development.

Table 18.1 Key questions for proving a neurodynamical theory

1.	How is intentionality communicated to the relevant action regions of the brain?
2.	How do humans learn?
3.	How does the human rapidly adapt to change?

Table 18.2 The minimum number of factors that must appear in a training program to instantiate KIII Theory (ARTT)

Factor	Relevance
Sports specific mesoscopic components	Key to the KIII theory
Component purpose	Intentionality driven
Experimentation	KIII implies individual initiative
Numerous samples	KIII implies successive approximations
ARTT	KIII implies rapid responsiveness to extremes
Stress/Fear	KIII implies rapid adaptation to change
Complex environment	KIII is chaos driven
Hands off approach	KIII implies self organization

After some analysis it was decided that the athletic analog of a mesoscopic component from [4, 6] must be the simplest possible motor action component that would have a purpose. Purpose was included to assure intentionality was present. This was necessary for validation of the theory as well as the fact that this was well supported by the research of Langer [8] at Harvard. We know from the research of Langer that motivation or intentionality would be an essential component for learning as well as for neurodynamics. This was consistent with the KIV theory, an extension of the KIII Theory to include intentionality [3]. The nature of tennis itself provided the answers we needed, see Table 18.3.

Each simple component was selected to be versatile, i.e., useful in serving more than one purpose.

In addition to the technical factors of the sport that required the derivation of mesoscopic components, we also had to consider two other factors: Physical conditioning and mental toughness. Without these additional components, no valid test of learning could be formulated since both conditioning and mental toughness can "trump" technical skill. This meant that no matter how well mesoscopic components were acquired by the student, their ability to use these skills in the sport depended on their conditioning and mental toughness. In fact, skill could be completely lost due to

Table 18.3 Learning components in tennis

Mesoscopic component	Component action
Primary component	Contact between the racquet and the ball
Secondary component 1	Extension through ball path
Secondary component 2	Rotation of racquet head into ball path alignment
Secondary component 3	Acceleration of racquet into ball path
Secondary component 4	Initiate racquet forward advance
Secondary component 5	Retract racquet into starting position

the extreme stress of formal match competition [2]. As a result of these known facts, we had to develop a program that supported skill development with physical conditioning and mental toughness training. This consideration required that we derive our program from the principles of eye-to-eye combat. There are three that apply to all combat from MMA to Tennis to war and they are derived from the single most important principle of combat: to break the enemies/opponent's will to continue, Table 18.4.

Table 18.4 Principles in breaking the opponent's will

1	Intimidate the opponent
2	Make the opponent feel physical pain
3	Inspire fear in the opponent with your assault

Fig. 18.1 Mesoscopic component analysis and conditioning program

To develop mental toughness we introduced the MMA training protocols of current bantam weight champion Ronda Rousey to further amplify the primitive demand on the mind and body. This included using MMA equipment such as body protectors to be able to include striking in the protocols. However, no head striking was allowed due to its risk of injury, see Fig. 18.1b. To assure that physical fitness was never a factor in skill performance, we developed an extensive physical fitness program, Fig. 18.1c. Each component of the physical fitness program had to be tied to mesoscopic component development. This assured that the exercise was relevant and driven by intentionality.

18.4 Example Results

Our research and analysis concluded that the instruction protocol would have to diverge drastically from legacy tennis training approaches. Most importantly, as our analysis from paleoanthropology shows, the role of language or explicit instruction would require significant attenuation. As a consequence, micromanagement of the student would have to be eliminated and replaced with a minimalist approach whereby the instructor was "nearly" a bystander. This element of the protocol also supported the non interference requirement necessary to assure minimal corruption of the data but also was consistent with the KIII Theory. Consistent with the mesoscopic wave-pulse approach would be the inference that once a component was developed, it would be used in a variety of contexts without the need for explicit direction [5]. This drove the protocol to depend on the individual's initiative and creativity to "fill-in-the-blanks" with minimal assistance from the instructor. Therefore experimentation and exploration by the student was a necessary ingredient of the protocol. Further, the derived training protocol must make maximum use of mimetic learning which can proceed at a remarkable pace [7] sometimes just a few minutes is all that is required. This is fully consistent with the mesoscopic wave-pulse theory and is also well-supported by the theory. Chaotic dynamics would be necessary as well. Chaos was included in several forms: constantly changing schedules; a random order of exercises and drills; surprise changes from projected schedules etc. The demand to adapt quickly is a consistent and necessary theme that must run through all training exercises and drills. The drills and exercises had to challenge the brain to develop and adapt due to the inherent difficulty of the program.

Within this complex environment, students from all walks of life were trained to play tennis and to execute even the most difficult and complex actions of the sport. Clearly, the demand to rapidly adapt to change in the face of fear and complexity is best explained by the mesoscopic wave-pulse theory. Another implication of this theory is that a part of the brain that originally starts out having a specific function could be reconfigured in a short time span to perform a different function based on the level of desire of the subject. A common example is that of the individual who, after losing the use of their arms, learn to use their feet for the original purpose for which the hands were used. An additional implication is that learning a sport must proceed from intense, nonverbal environmental activity [1]; further, the process of trial-and-error must be permitted [8]. One significance of including trial-and-error is that it allows for the brain to "self organize" over very short time spans. Short-term self organization is consistent with the KIII wave theory but not with the static neuronal theory.

While our approach is to train normal subjects using the KIII protocol, having functionally normal subjects with some accidental or embedded abnormality would be most useful. Using this approach, a simple test of the KIII hypothesis could be conducted if a subject entered our program with a known physical limitation. This had to occur by chance since proselytizing and advertising were prohibited. As chance would have it over a 16 year period, players did arrive on our door step who had lost some range of motor action. We will only mentioned three here.

Case 1 The first medical case is of a subject who entered our program with an inoperable tumor on his brain stem. See Fig. 18.2b which includes the MRI of Dan T.

Fig. 18.2 Experimental results

dated 2011; and, the figure also shows Dan T. is performing effectively in competition. His prognosis was not fatal, but that his ability to engage in complex athletics would likely never be possible. At one point, the mother of Dan T. was told not to remove Dan from hospital care because he would die. She ignored the doctor's advice. By chance Dan T's family was a member of the Plaza Oaks Club in Houston, TX out of which this tennis academy operated. She brought him, and his younger brother to Jana van der Walt, our head pro to begin lessons. Within two years Dan T was defying all the odds and learning to play tennis well enough to compete in formal sanctioned tournaments. In his second match, played in December of 2014, Dan T took his match to a third set tie-breaker against a far superior normal student.

Dan T's starting point in our program requires mentioning. When Jana began feeding Dan balls that were high, he would duck; on some occasions the ball would hit Dan as he was unable to make decisions about how to adjust to the ball path. The significance of this is that in Fig. 18.2b, Dan is hitting a ball out of the air in a formally sanctioned USTA tournament. This photo demonstrated that not only had Dan formed the mesoscopic components of movement to adjust to, and track the ball in an extraordinarily short time span, he did it under the pressure of competition.

Serendipitously, two important points emerged from the fact that Dan T was trained independently by Ms. Walt without any advice beyond following the KIII training protocol: (1) the method itself was demonstrated to be transferable; and, (2) that implementing the KIII Theory in a sports venue may be independently verifiable.

Case 2 A second example is that of Jean V. See Fig. 18.2a. Jean V destroyed his knee in motor cross training. The exact diagnosis from the MRI of the left knee joint was: *Acute full thickness versus near complete mid substance tear of the posterior cruciate ligament. Anterior cruciate ligament is intact. High grade sprain of the of the proximal superficial medial collateral ligament. Moderate tear of the medial collateral ligament. There may be a complete tear at the femoral origin. Oblique tear within the middle and peripheral thirds of the posterior horn of the medial meniscus.* The rest of the knee was found intact.

Jean V's injury was a result of a fall while involved in motor cross training. Surgery was recommended, however, Jean V elected to use our program as a rehabilitation venue on the premise that his brain would, with extreme intentionality and determination, shortly reroute his muscle groups to allow him to return to playing competitive tennis. Full dynamic recovery occurred within less than three months. In addition to Fig. 18.2a, Jean can be seen grappling in the MMA figure, Fig. 18.1b, bottom right.

The foregoing examples are of students who had some limitation that was readily overcome by the KIII teaching methodology. The following examples are two students who have no limitations and were candidates for professional tennis. However, the student's parents thought that they could do even better and moved the student to another program. This provided us with an excellent opportunity to make a direct comparison of KIII theory development with conventional methods. A key to understanding the figures is the RPI, or Relative Performance Index. This is a number that is calculated each week by IMG for all amateur sports. It measures the candidates performance relative to their competitors. This is the only relevant decision making factor that is a pure number that is important to scholarship decisions or other coaching decisions.

In Fig. 18.2d, we show the RPI of the KIII theory beginning around September 2013–April 2014. At this time the parents moved the student to a conventional program after a short transition period. The conventional program started in September 2014 and is still in progress at the time of this publication. In addition to this case, there have been several other cases where, for various reasons, the student had to transition to a conventional program with the same relative RPI results.

Of particular significance is that during the student's KIII training period we see steady improvement well beyond the expected. On the other hand, when the student returned to a conventional program, the RPI flat-lined. In Fig. 18.2c, we show the RPI of for a short time duration student in the KIII theory beginning around June 2014 to the beginning of August 2014. At this time the parents, as with RH, moved the student to a conventional program. The conventional program started in August 2014 without a transition period and is still in progress at the time of this publication. After examination of several more cases, we are predicting that the KIII program will out perform conventional training by a significant margin. This prediction is illustrated in the bottom right of Fig. 18.2c where we predict that the conventional program will reflect slow but steady progress whereas the KIII program will produce very rapid progress. If these results can be replicated in rehabilitation programs, this would mean that a rehabilitation program derived from the KIII theory could result in very significant improvements in every form of rehabilitation.

18.5 Summary

In this paper we have demonstrated how to instantiate the KIII Theory in a tennis training venue which is a very simple model of a human system involving competition, stress, rapid development and complex decision making. The objective results on the USTA tournament Websites demonstrate that students who have come through this system compete very effectively against students that have been trained in conventional protocols and that even students with limitations have developed far faster than is possible using other systems. This would suggest that the KIII Mesoscopic Wave-pulse Theory provides a very effective training protocol when translated into practice.

Acknowledgments While many students passed through our program over the years, we would like to acknowledge those who can be clearly identified with our program. To do this we confine our acknowledgments to those who were in the program at least two years; or, based their development directly on using the KIII Theory protocols; and, who are present in our program as of December 30, 2014.
Students Hudson Lorfing, Jean Valiquette, Vishal Puppala, Anuj Samal, Mircea Tatulescu, Dan Tatulescu, Kristen Huggins, Faysal Alameddine, Aya Alameddine, Claudia Mollerup-Madsen, Ghida Zaatari, Vijay Patel.

Assistants Jana van der Walt, Mari van der Walt, Henry Garza, Jason Haynes, Shikha Uberoi
Financial Support Ricardo Nazario, W. C., Dr. Dileep Puppala, Denise Slaughter, Diane Zvera, Dr. G.S. Ramesh, Dr. Yamini Naygandhi, Dr. Reka Ramesh, Dr. Fadi Alameddine.

References

1. Campbell SB (1995) Behavior problems in preschool children: a review of recent research. J Child Psychol Psychiatry Allied Discip 36(1):113–149
2. Freeman WJ (1995) Societies of brains. Lawrence Erlbaum Associates, Hove
3. Freeman WJ (1999) How brains make up their minds. Weidenfeld & Nicholson, London
4. Freeman WJ (2000) Neurodynamics: an exploration in mesoscopic brain dynamics. Springer, New York
5. Hippocrates Adams F (1939) The genuine works of Hippocrates, Kessinger legacy reprints. The Williams & Wilkins Company, Baltimore
6. Ilin R, Kozma R (2006) Stability of coupled excitatory-inhibitory neural populations and applications to control of multi-stable systems. Phys Lett A 360:66–83
7. Kandel E, Schwartz J, Jessell T (2000) Principles of neural science, 4th edn. McGraw Hill, New York
8. Langer E (1985) The power of mindful learning. Addison-Wesley, Reading

Part V
Commentaries on New Theories of Cortical Dynamics and Cognition

Chapter 19
Commentary by Paul J. Werbos

How Can We Ever Understand How the Brain Works?

Paul J. Werbos

19.1 Introduction

The preface to this book raises several questions which are important to our understanding of the brain, in specific terms, and to the larger question of the strategy we use to try to develop a more complete understanding in the future:

1. How can we explain Freeman's empirical observation that the cerebral cortex regularly undergoes abrupt shifts, following some mix of the alpha and beta rhythms, above and beyond the higher frequency components which he like many others has also observed?
2. Are the kinds of data he has collected, at an aggregate systems level, of any real value as part of the effort to understand how the brain works?
3. In general, is bottom-up modeling, constrained by the many assumptions of today's varieties of "neuron dogma", the only important stream of the path which could take us to a mathematical understanding of the underlying intelligence of the brain?
4. Is it possible that field effects, above and beyond the what is usually modeled in dendritic compartmental models or abstractions of the same, could play a fundamental role in enabling the kinds of minds and intelligence we actually see in nature, and in opening the path to constructing more intelligent computational systems?

Though I deeply value my ongoing collaborations with Freeman and Kozma (and past joint ventures with Pribram and Pellionisz), my views on these questions are not identical with theirs. Thus I am especially grateful to them for offering me a chance to review my own views on these questions. This chapter will begin by discussing questions 2 and 3, the methodology issues, as part of the larger strategic question of what we all should be doing in research, in order to move as directly and swiftly

P.J. Werbos
Department of Mathematical Sciences, University of Memphis, Memphis, TN, USA

R. Kozma and W.J. Freeman, *Cognitive Phase Transitions in the Cerebral Cortex – Enhancing the Neuron Doctrine by Modeling Neural Fields*, Studies in Systems, Decision and Control 39, DOI 10.1007/978-3-319-24406-8_19

as possible to a true functional, mathematical understanding of intelligence in the brain—intelligence and mind, the phenomenon which many of us consider most important here. Sections 19.3 and 19.4 will then give my views on the more concrete questions, 1 and 4, in that order.

19.2 Top Down Versus Bottom up and the Neuron Dogma

As cognitive science and brain science begin to give us real insights into our own intelligence, they can even give us useful insights into the conduct of neuroscience itself. For example, Simon's group did important studies at CMU decades ago on problem-solving strategies by humans and machines. He found that many people use "forwards induction" strategies, trying to do something good to take them closer to a goal, hoping that bit by bit they may finally get there. Many of the bottom up strategies of normal science are like that. For some problems, forwards induction can eventually work. But for ill-structured problems in very complex spaces, requiring adaptation, it is essential to include an element of "backwards induction" to have much chance of success. Backwards induction entails clear visualization of a goal, and thinking backwards about how to get to that goal, with a lot of strategic thinking involved. Trying to really understand the brain and the mind, with the full richness of mathematical science, is one of the most complex challenges ever faced by humans; thus I would claim that there is very little chance of ultimate success if we do not formulate clear longer-term goals. That is why it is essential that we break off from the details at times, with some regularity, to review our general strategy here. This is also an important part of what we need to do, to interface with issues of funding, where goals, milestones and ultimate outcomes are always in order.

Many leaders of science funding now recognize the importance of funding new cross-disciplinary efforts focused on larger societal problems such as lower cost large scale sustainable energy. They recognize that concrete strategic thinking and new cross-disciplinary teams are essential to progress in those areas. (Of course, when those challenges are difficult, effective strategic thinking requires an adaptive approach fully informed by concepts of decision trees and risk management [10, 12].) I have argued in many forums that the most important grand challenges of fundamental science, like understanding the brain, also require more of that kind of approach, if we really want to get to the goal.

I am grateful that the Capitol Science Forum of the Washington Academy of Sciences invited me to give a keynote talk on this subject in 2012. In that talk, I pointed to the NSF one-time initiative on Cognitive Optimization and Prediction [3] as a model for the new kinds of cross-disciplinary activity we should be supporting. I reminded people that the universe of possible nonlinear mathematical models which could POSSIBLY be applied to the brain is truly immense, and that it may be grossly inefficient to model the brain as if it were just another random complex system in nature, like a collection of clouds or the ecology of a scummy pond. In Grossberg's terms—if we are trying to filter through a very huge space of models, and find "the right one", we really need to make full use of all the filters available to us, in an integrated way, in order to locate models which are consistent with all three requirements—fit to neuroscience information, fit to psychology experiments, and fit to our knowledge about the ability of brains in general to learn to solve extremely complex general types of problems. COPN did not displace existing efforts within narrow disciplines, but it supported new cross-disciplinary groups to carry this forward. The acting director of NSF was in attendance, and it seemed for awhile that there might be hope of a new focused national effort to move us more quickly to better true functional understanding of the brain.

To apply backwards induction effectively, we do need to create some crisp and workable statements of goals. These kinds of specific grand challenges should not displace all other science, but including some real focused energy on key grand challenges is essential here. In COPN, we basically posed the following grand challenge: how to understand and replicate that level of general intelligence possessed by even the smallest mammal brain, situating it in the staircase of levels of general intelligence in the vertebrate lineage. (We understand that the great lineages leading to ants and bees, and to octopi and their relatives, are also very interesting in principle, but not as important to the specific goal of replicating and understanding the higher intelligence of the mouse.) To make the goal better defined mathematically, we focused on two specific learning abilities: the ability to "optimize", to perform better and better in tasks requiring decision and control, and the ability to "predict", which entails building up inner models of the environment to predict, anticipate, filter and infer things about things unseen in the environment.

COPN was led by the Engineering Directorate of NSF. For engineers trying to mimic the general learning capabilities of the brain, with artificial neural networks (ANN) or other biologically inspired methods, it is extremely discouraging that many of the models popular in computational neuroscience do not translate well into the mathematics of systems which make sense in performing general optimization or prediction tasks, or into the requirements of serious industrial-grade challenges. In order to find models of the brain which are truly functional, we need a better fundamental understanding of the larger class of models of which such models would be a subset: dynamic distributed systems capable of really effective performance in such learning tasks. We need crossdisciplinary collaboration to try to understand the relation between the most powerful models in that class (as in COPN [3]) but we also need core work by algorithm developers in systems engineering and related fields, to improve our knowledge of that class of mathematical system in general. In 2014, at the International Joint Conference on Neural Networks, I provided a more concrete review of where we stand today in developing the required mathematics, and a roadmap built on specific new tools which can take us all the way to the mouse level [18]. Based on my experience of how long it takes our communities to address the kinds of subgoals given in that paper, I would estimate that a new focused effort really could get us to the point of truly replicating the higher intelligence of the mouse in about a hundred years. In my view, that is the one the two really great, attainable grand challenges before all of science in the coming century.

But will we really move that fast on earth? At the present time, I am very worried. Once it became known that there might be money on the table, and the usual Washington lobby/stakeholder/PAC system came into effect, a report came out extolling the goal of getting to a true functional understanding of intelligence in the brain—and proposing billions of dollars focused just on expanding various types of brain mapping, as has been proposed ever so often in the past, with lip service to the idea that someone might find some idea someday for how best to actually use whatever data the technology developers might enjoy developing. This causes me to imagine a cartoon, in which the British government at the time of the young Charles Darwin offers to help him in what he plans to do, by spending billions of pounds on leaf collection and taxonomy, and building a new iron triangle which would actually exclude the kind of work Darwin went out later to do. Data are important, but I like the COPN model of designing new data collection as part of a more substantive scientific strategy wherever possible. (But again, this is not to criticize existing archiving efforts.) I am especially worried about potential damage to human guinea pigs in any effort which does not pay enough attention to the goal of understanding the brain, or to accounting well enough for what we already know about the risks of short-circuiting reinforcement networks.

Some technology hardware people have even claimed that we are already approaching the "cat level" in artificial brains. It is important to remember that this refers to hardware implementation of specific models from computational neuroscience. It was a telling moment, in my view, when HP pulled out of the DARPA SYNAPSE program, because of a need for greater flexibility, to develop new hardware architectures with more hope of actually including some in the class which would actually work, from an engineering perspective [7]. Our ability to develop ever more powerful hardware is extremely noteworthy, but without sophisticated enough algorithms and architectures (and flexibility to support it) we do not really have a brain. We need to work hard to learn more about brains and about the relevant mathematics to have any hope of this.

To answer questions 2 of the introduction, I would refer back to this context. Many people are now working on the crossdisciplinary task of developing the kind of mathematics and algorithms we need in order replicate mouse-level capabilities [9, 18], but I personally am one of the relatively few people in that group deeply involved in the development of the new functional mathematics who has worked hard to learn ever more from neuroscience, and take the COPN approach in my own work. (Of course, I wish for funding to develop many more, and COPN itself was a good if small start.) As in the COPN announcement, I have found it especially valuable to work with SYSTEMS neuroscientists, like Pribram and Freeman, in order to get the basic systems-level understanding necessary before one can really understand the function of important subsystems of the brain, let alone how they accomplish those functions. When tensions between cellular-level people and systems level people get in the way of our learning, the most we can from both levels, it interferes with progress. That is just one example of disciplinary or tribal loyalties and fragmentation getting seriously in the way of what needs to be done, to have any hope of getting to the larger goal. It makes me wonder what life would be like in a small town, if all the grocery shop owners decided that only grocer shop owners are worthy of being fed, and refused to sell food to anyone else, including even the farmers who supply them. We who study how networks can give rise to intelligence should be more self-conscious about building our own networks of science, and avoiding subconscious decisions based on false ideas, like the idea that intelligence is greatest when all neurons are synchronized to output the same value ("1" or "0") in unison.

In the next section, I will address question 1 of the introduction, which gets into the specifics of what we are seeing with EEcog and multichannel recording. I do believe that data collected by people such as Barry Richmond, Nicolelis and Chapin, and Jennie Si with multichannel unit recording do have a lot of information vital to systems-level understanding above and beyond what can by learned from EEG and EEcog alone. However, I commend Freeman here for getting really explicit about what he sees of importance to the functional types of inquiry and modeling. Many of the most exciting results of Richmond, and of Nicolelis and Chapin, are very hard to see in the actual published papers. Perhaps if they had been as blunt (and as chaos-oriented?) as Freeman, and if the journals had let them get away with that, the novel information content of this new work would be less than it is, but perhaps not.

19.3 An Approach to Explaining the 4–8 Hertz Abrupt Shifts in Cortex

Freeman notes that he sees abrupt shifts across wide regions of the cortex, at a rate fitting the alpha or theta rhythm or a mix of the two. He rightly notes that the usual type of asynchronous model, such as ODE models or spiking neuron models, would normally not predict this kind of behavior, with this degree of repeatability. What could explain such shifts?

One possibility could be that they are caused by field effects beyond the scope of a wide class of neural network models. While I agree that field effects may well be important to the brain and the mind, I do not see them as the most promising approach to address this specific question. In essence, it passes on the problem of explaining the abrupt shocks to another level, where many questions still need to be answered.

Personally, I would advocate following up on a second type of new explanation which follows from my own theory of how the cerebral cortex works [15–17]. Sometimes, when there is a high degree of regularity in something which affects all neurons in the cortex, the effect may be due to something which is hardwired in nature. In my view, the behavior of the giant pyramid cells which dominate the cerebral cortex is driven in turn by timing or "clock" signals sent from the nonspecific thalamus to a diffuse network of axons which synapse directly on the waistline at the base of the apical dendrite. These clock signals switch the relation between the apical part of the neuron and the soma part, so as to implement a prediction algorithm which fits well with Freeman's term "cinematic processing".

More precisely, the theory asserts that the predictive aspect of the cortex implements something very similar to the SEDP design given in the Handbook of Intelligent Control [23] (Chap. 13), with the soma calculating $\tilde{R}$ and the apical dendrite $\hat{R}$, and both being hybrid time-lagged/simultaneous recurrent networks [23] (Chap. 13).[1] In designing neural network systems which actually work, it is important to understand the distinction between time-lagged recurrence (essential for short-term memory and learning of causality in observing dynamical systems) and simultaneous recurrence (essential for more powerful pattern recognition capability). The hybrid networks or time-lagged recurrent networks (TLRN) which work well in engineering would also show abrupt changes when the clock pulses come in which manage the time lags in a way which also allows short-term memory and learning of causality. Believers in simple, totally asynchronous neural networks find it hard to imagine that even brains require this kind of extra complexity to do their job, yet Freeman's observations suggest that nature might be quite similar after all.

Of course, this is all just one aspect of the theory, and it is important to consider other streams of empirical evidence as well. See [15–17] for more details, and [3] for some hints about additional experiments to further test the theory.

This theory does not really say that ODE models cannot replicate the abrupt transitions, if we construct ODE which match the simpler systems level theory I have referred to. One could model the pacemaker cells in the nonspecific thalamus as hardwired oscillators, not unlike cardiac pacemaker cells, and invoke strong nonlinearity at their synapse with the apical dendrite, to model the actual physics of how the discrete clocking and switching is implemented. This would be similar in a way to the circuit model differential equations used to explain how simple "not" elements in a digital circuit actually come to behave as the design calls for them to behave, as simple "not" elements. Both levels of description have their place, and relate to each other, but the "not" level lumped description is perhaps better as a way of describing the information level of a digital computer under normal conditions.

Several years ago, Rodolfo Llinas gave an impassioned plenary speech at the International Joint Conference on Neural Networks, reviewing how pervasive and precise the "clocks" in the brain really are, and how asynchronous models which do not reflect their importance are simply not plausible. This all fits nicely with Richmond's findings on the neural code, discussed further in [15–17].

Again, this explanation for the transitions is not an argument either for or against field effects in the brain. Time-lagged recurrence, simultaneous recurrence and even clocks are mathematical concepts which can apply equally to discrete systems of system variables (as in traditional neural network models) or to systems made up in part of field variables.

---

[1] Both chapters are posted at URL: www.werbos.com/Mind.htm mouse, which is legal because they were written by a government employee on government time.

19.4 Could Field Effects Be Important to Brain and Mind?

Could field effects be important to an understanding of intelligence in the brain and mind, to such an extent that they need to be represented explicitly at the systems level in order to fully replicate that intelligence?

This question pushes the limits of our knowledge. Our new mathematical theory still leaves room as yet for alternative possibilities here. Thus the scientific method calls for us, not to form opinions, but to look for ways to evaluate the possibilities.

For myself, I am interested in three ways in which field effects might possibly be so important:

1. Theories of associative memory, or even quantum associative memory or even other forms of local quantum computing (e.g. stochastic optimization) within the cytoskeleton of the neuron;
2. Concepts of dendritic field processing [14], articulating key ideas of Karl Pribram in resonance with much of the writing of Freeman as well, based on lengthy empirical observation;
3. True quantum information processing at the systems level. This discussion will discuss these three possibilities in that order.

19.4.1 Associate Memory or Quantum Effects Inside the Neuron

Many researchers through the years have argued that the simple McCullough-Pitts model of the neuron (even with memory and feedback effects and a bit more nonlinearity added) is much too simple to reflect the full processing capabilities of an individual neuron. Speakers at Hameroff's 1991 NATO workshop on the cytoskeleton showed videos of how the amoeba—another single cell information processor—can learn to perform very complex trajectory calculations, to allow it to "zap" a heat source within its range. They argued that the neuron, a cell which is even more specialized to assist complex information processing, must have evolved even greater complexity in what it can do. Skeptics argued that amoebae are more like free humans, whilke neurons are more like soldiers in an army, and that armies have been known to dumb down the repertoires of their members for the sake of order. For decades, many researchers have argued passionately that systems such as the microtubules could be performing an associative memory function. Michael Conrad, an early leader of the molecular computing community, developed models of how stochastic optimization might be implemented and used within an individual neuron.

In the discussions leading to COPN [3], three main possibilities were discussed:

1. The classic "neuron dogma", which simply dismisses these possibilities;

2. The associative memory concept, in which individual neurons are claimed to perform normal input/output mappings like those of an associative memory, far more complex than anything like a McCullough model;
3. An intermediate model, where the input/operation of the neuron is like the McCullough-Pitts neuron but complex associative memory effects or matrix processing are used to support learning within the neutron, similar in spirit to the EKF learning which Kozma and I have used on occasion [5].

To get past speculative debates, COPN called for researchers interested in (1) to try to propose experiments in which individual neurons would be trained to learn input/output relations beyond the capabilities of a simple McCullouhg-Pitts neuron, such as perhaps the XOR mapping to begin. No proposals were received on that topic as such, but two of the four big proposals funded by COPN did include a strong focus on training neural systems in culture. In the course, of that work, we learned that much of the work in cultured neural systems assumes an asynchronous neural code, and does not include the kinds of timing signals which I would consider crucial to success here. Ironically, work to transfer such methods to use in the brain did include such timing signals, because many people working with whole brains understood their central importance, but the technology of communicating with neurons in culture and motivating them to be trainable is still at any early stage. Such research gives us an essential opportunity to learn more clearly what attributes of the whole brain are important to what levels of learning, but the work does not yet seem to be at a stage where it can help resolve the issue of (1) versus (2) versus (3). Other approaches may well be possible, but I have not seen specific suggestions for how.

19.4.2 Dendritic Field Processing

Many years ago, Karl Pribram presented a graphic story of how dendritic field effects might bring capabilities to the brain greater than those of simple McCullough-Pitts neurons acting in isolation. He proposed that there are layers in the dendritic fields, where the linear field interactions still enable a kind of complex matrix operation across cells, leading to subsystem capabilities like those of Hopfield networks.

In one of his workshops and books [14]. I did my best to try to translate his story into a new kind of dendritic neuron (or neural layer) model which might well allow capabilities greater than that of a traditional neural network, understandable and even perhaps useful in engineering as well. There was some later work on piecewise or partially linear subsystems in learning networks (e.g. by Leen and by Atkeson) which gave some hints about the possibilities here, but on the whole, the subject was simply not explored systematically because those who could have taken it further had too many other things to do. Logically, this topic is one more question within the realm of the larger program of "vector intelligence" [18], for which considerably more research is needed. At one of the IEEE conferences on Systems, Man and Cybernetics, I briefly discussed the research issues in how to integrate this kind of processing with other elements we know are needed for full, optimal vector intelligence.

19.4.3 Quantum Fields and Quantum Mind

In Sect. 19.2, I argued that understanding and replicating the level of intelligence we see even in the brain of the smallest mouse is one of the greatest grand challenges to science in the coming century. The other great grand challenge, in my view, is to nail down the "theory of everything", by cleaning up our understanding of quantum field theory and of the three basic types of force (electromagnetism, nuclear, gravity) we have begun to understand.

In my view, having probed the limits of what we really know about quantum computing and the quantum grand challenge, I see very little possibility that quantum computing effects could be relevant at a systems level to understanding the intelligence of an isolated mouse brain, the focus of Sect. 19.2. Yes, there is life beyond the mouse, but we do not yet have the prerequisites to understanding that life on a fully scientific basis, within the kinds of constraints reviewed by Thomas Kuhn [8]. As humans, we have good reasons to ask about life beyond the mouse, and to try to develop our own very highest mental capabilities (as individuals or as part of larger networks), but we all need to maintain lots of caveats and tolerance of diversity in the venues we have for such exploration. In my view, the effort to understand the mathematics of intelligence up to the level of the mouse already gives us a clear understanding of the core ideas of Freud on the subject of psychodynamics [17], but we still have a lot of work to do to do full justice to Jung, let alone understanding the balance between collective intelligence effects and more worrisome effects such as those described by Spengler and those we see related to changes in the flows of funds feeding back to political and cultural policies.

Still, there is an emerging scientific basis for understanding how a kind of quantum level of intelligence is possible, relative to which our individual human minds would not be truly "conscious" [19].

In pursuit of the physics grand challenge, I recently encountered some new results which have substantially changed my own understanding of "where mind is located" in the physics of the universe [21]. Like most scientists who fully understand Western traditions, I have always tried to understand mind as a kind of pattern which emerges as a form of organization of the fundamental substance or physical substratum of our universe. This paradigm is far more powerful and open than most naïve mundane thinkers imagine. I was also very excited this past year to see more and more evidence that we can formulate the "law of everything" in a realistic way, assuming a cosmos which is ultimately based on three dimensions of space and ne of time, or something very to that. I had deep respect for the work of David Deutsch, the true father of today's version of quantum computing [1], but I was very skeptical of his theory [2] that we live in just one time track within an infinite multitude of time tracks, interfering and leaking with each other in a way which classical brains would find it hard to track.

More precisely, I have been interested in a class of theories of physics built around what might be called scenarios or paths, similar to those of classical field theory. In classical Hamiltonian field theory, the state of the entire cosmos over all space-time is specified by specifying two vector fields, $\underline{\varphi}(\underline{x}, t)$ and $\underline{\pi}(\underline{x}, t)$, at all points $\underline{x}$ in space and all times t, where $\underline{\varphi}$ and $\underline{\pi}$ each may have many components such as objects which would be called tensors in relativity theory. Each such state S over space-time may be called a "path" or a "scenario". Both in standard Feynman path physics, and in stochastic realism, the laws of physics are defined by models which predict the probabilities of alternative scenarios S.

Nevertheless, as I looked more closely at the details of how stochastic realism describes some of the basic experiments with entangled photons, I was startled by what I saw. As I looked very closely at the motion of a polarized photon traveling through a polaroid polarizer, and considered my own motion through the polarizing environment of Washington DC, I understood how the kinds of mathematical complexity and form we see as intelligence is actually a set of patterns within *POSSIBLE* tracks within the simpler real universe. All the complexity of mind appears to exist within the scenarios themselves, and not just in the final outcome. In effect, the sudden shock was the feeling "We have found reality, but we are not it". The intuitive content of Deutsch's theory, and his view of the place of individual minds within the universe, does appear to hold up, even when the mathematics is simplified as far as we can push it. What this really means for the highest level of intelligence in our universe is something I am only now just beginning to understand.

The mathematical issues related to "where is mind" have a close relation to earlier work by Freeman, Kozma and myself [4]. Freeman and Kozma previously did extensive simulation studies of neural network models based on traditional nonlinear dynamical systems, which still somehow did not capture key aspects of the emergent behavior seen in the areas of the brain being modelled. Thus they added a small amount of noise to the model, which could then be expressed in general terms as:

$$\dot{\underline{x}} = \underline{f}(\underline{x}, W) + c\underline{u} \tag{19.1}$$

where the vector $\underline{x}$ is the set of variables in the model, W is the set of parameters, and $\underline{u}$ is a vector of random noise. The remarkable finding, in simulation, was that the emergent properties of this model would be very different from those of the model without that term, even in the limit as c goes to zero. We called this phenomenon stochastic chaos. A stochastic chaos model was able to replicate the observed dynamical behavior of the actual system. Related work in mathematics proved rigorously that stochastic chaos does exist from systems $\underline{f}$ based on quadratic mappings [6].

This leads to the question: when are the emergent properties of stochastic realism models (which are essentially just Hamiltonian field theories with a time-symmetric noise term added) qualitatively different from those of the corresponding Hamiltonian field theory, even in the limit as c goes to zero? Would the emergent properties of mind be like what David Deutsch portrays in both cases [2]?

Again, these questions emerged when I thought deeply about what I actually see in a new model to predict the motion of one photon in a polaroid polarizer, as part of a model which successfully predicts the standard Bell's Theorem experiments without the old metaphysical assumptions called "collapse of the wave function" [20, 22]. Truly, in watching that photon intently in my mind, I felt a bit like Alice in

Wonderland, following a white rabbit who propelled her unexpectedly into a totally new world. In retrospect, however, the same complexities apply, with even more force, in the older form of quantum field theory (QFT) which has become holy writ to many high-energy physicists, the Feynman path formulation [13].

19.5 Summary and Conclusions

This book contains a chapter by Vitiello [11], who stresses the need to develop dynamical systems models of the various types of brains. It is easy to be hypnotized by the complexity of bottom-up information about the brain, but the complexity of the physical universe itself is greater than the complexity of the brain. Physics became a coherent, strong branch of mathematical science after Isaac Newton firmly refocused attention away from that complexity to the study of the *laws of change*, which are more universal and easier to comprehend. This chapter has argued that we now have a specific roadmap before us to achieve a similar Newtonian revolution, by extracting and reverse-engineering the principles of learning which underlie the highest capabilities of the mouse brain (and of its ancestors).

Do we need to invoke the full complexity of many-body quantum electrodynamics (QED) to achieve that revolution? This chapter has argued that we should remain open-minded about that question, and focus our energy on developing a better mathematical understanding of the types of dynamical system which may support powerful intelligence.

Vitiello's [11] quotation "the brain is not a stupid star" is important at many levels. Above all, when we search for models of intelligent brains (classical or quantum), we need to remember that the brain is not just a random point in the space of possible dynamical systems. It is a set of points in a much smaller space, the space of dynamical systems which have truly unique functional capabilities, capabilities more like those sought in advanced engineering design than those we find by making analogy to dead systems in nature. The functional approach is the key to achieving the same kind of thing in neuroscience which Newton achieved in physics.

With brains, there is an additional methodological trap we always need to be aware of. The dynamical equations which describe what any individual brain has learned in a lifetime will also be very complex, and overwhelming. It is the dynamics of learning as such, which govern that complexity over time, which offer hope of a Newtonian kind of useful simplification and universal understanding. Whatever the limits of such understanding, it is more valuable by nature than simply drowning in detail.

The human mind and our larger society are much larger than a simple isolated mouse brain, but a deep and full understanding of the latter can play an essential role in helping us outgrow at least the most extreme delusions and dogmatic theories about the former, and achieving a better understanding of human potential [17].

References

1. Deutsch D (1985) Quantum theory, the Church-Turing principle and the universal quantum computer. Proc R Soc Lond Math Phys Sci 400:97–117
2. Deutsch D (1997) The fabric of reality: the science of parallel universes and its implications. Penguin, New York
3. Emerging Frontiers in Research and Innovation (2008) National science foundation, Arlington, VA. http://www.nsf.gov/pubs/2007/nsf07579/nsf07579.pdf
4. Freeman WJ, Kozma R, Werbos PJ (2001) Biocomplexity: adaptive behavior in complex stochastic dynamical systems. Biosystems 59(2):109–123
5. Ilin R, Kozma R, Werbos PJ (2008) Beyond backpropagation and feedforward models: a practical training tool for more efficient universal approximator. IEEE Trans Neural Netw 19(3):929–937
6. Kozma RT, Devaney RL (2014) Julia sets converging to filled quadratic Julia sets. Ergod Theory Dyn Syst 34(1):171–184
7. Kozma R, Pino R, Pazienza G (eds) (2012) Advances in neuromorphic memristor science and applications. Springer, Berlin
8. Kuhn T (1996) The structure of scientific revolutions, 3rd edn. University of Chicago Press, Chicago
9. Lewis FL, Liu D (2012) Reinforcement learning and approximate dynamic programming for feedback control. Wiley, New York
10. Raiffa H (1968) Decision analysis. Addison-Wesley, Reading
11. Vitiello G (2015) Filling the gap between neuronal activity and macroscopic functional brain behavior. doi:10.1007/978-3-319-24406-8_22
12. von Neumann J, Morgenstern O (1953) The theory of games and economic behavior. Princeton University Press, Princeton
13. Weinberg S (1995) The quantum theory of fields. Cambridge University Press, Cambridge
14. Werbos PJ (1993) Quantum theory and neural systems: alternative approaches and a new design. In: Pribram K (ed) Rethinking neural networks: quantum fields and biological evidence. Erlbaum, Hillsdale
15. Werbos PJ (2009) Intelligence in the brain: a theory of how it works and how to build it. Neural Netw 22(3):200–212
16. Werbos PJ (2010) Mathematical foundations of prediction under complexity. Erdos Lectures/Conference 2010, Memphis, USA, http://www.werbos.com/Neural/Erdos_final.pdf
17. Werbos PJ (2012) Neural networks and the experience and cultivation of mind. Neural Netw 32:86–95
18. Werbos PJ (2014) From ADP to the brain: foundations, roadmap, challenges and research priorities. In: Proceedings of international conference on neural networks IEEE. http://arxiv.org/abs/1404.0554
19. Werbos PJ (2014) Time-symmetric physics: a radical approach to the decoherence problem. In: Steck J, Behrman E (eds) A radical approach to the decoherence problem. In: Proceedings of the workshop on quantum computing of the conference pacific rim artificial intelligence (PIRCAI)
20. Werbos PJ (2015) Analog quantum computing (AQC) by revisiting the underlying physics. In: SPIE Proceedings of the quantum information and computation XIII, SPIE 9500-53
21. Werbos PJ (2015) Links between consciousness and the physics of time. Int IFNA-ANS J. www.werbos.com/Mind_Time.pdf
22. Werbos PJ (2015) Stochastic path model of polaroid polarizer for bell's theorem and triphoton experiments. Int J Bifurc Chaos 25:1550046
23. White DA, Sofge DA (eds) (1992) Handbook of intelligent control. Van Nostrand, New York

Chapter 20
Commentary by Ichiro Tsuda

Self-organization of the Second Kind: A Variational Approach

Ichiro Tsuda

Abstract **Devil's Advocate** This paper addresses a provocative remark advocating the field theory of Kozma and Freeman, by proposing a new mathematical model for functional differentiation in neural systems. The computational result suggests the presence of self-organization of the second kind.

20.1 Self-organization and Field Theory

Studies of self-organization have a long history, starting with the epoch-making movement of cybernetics and control theory [11]. Among others, Nicolis and Prigogine, and Haken [7, 10], and their colleagues and followers dramatically developed theories of self-organization and extended their applications to elucidate the mechanisms of pattern formations at macroscopic scales in far-from-equilibrium systems. Prigogine et al. and Haken et al. established such theories using the concepts of dissipative structure and entropy production and the slaving principle, respectively. These theories have also been applied to large-scale neural systems in the brain, for example, the appearance of oscillatory behaviors in a thalamo-cortical interaction pathway in terms of populations of excitatory and inhibitory neurons [1, 22]. One of the important characteristics of the self-organization theory lies in the concept that the cooperative interactions of atoms and/or molecules at the microscopic level bring about order formation at the macroscopic level, and that such ordered patterns govern the microscopic dynamics. For example, taking ephaptic couplings of neurons into account, it is necessary to reformulate the neuronal activity in the framework of field theory, based on data measured by Electroencephalograph (EEG), Electrocorticograph (ECoG), Magnetoencephalograph (MEG), etc. [3, 4, 14]. Because far-from-equilibrium states are maintained as stationary states only in the presence of energy dissipation, far-from-equilibrium systems are necessarily dissipative systems; thus, environmental variables are not involved in the systems variables, and they remain

I. Tsuda
Research Institute for Electronic Science, Hokkaido University, Sapporo, Japan
e-mail: tsuda@math.sci.hokudai.ac.jp

R. Kozma and W.J. Freeman, *Cognitive Phase Transitions in the Cerebral Cortex – Enhancing the Neuron Doctrine by Modeling Neural Fields*, Studies in Systems, Decision and Control 39, DOI 10.1007/978-3-319-24406-8_20

only as thermal noise that is associated with a dissipated energy, or are reformulated as control parameters. Freeman and Kozma (see, for example, [3]) and Tsuda (see, for example, [16]) tried to incorporate environmental variables in the system as external noise, and thus an overall system of interest should be a nonautonomous dynamical system formulated by a random dynamical system [12]. By keeping control parameters at fixed values, a dynamical system describing the averaged motion of the system represents the motion of order parameters, i.e., pattern formation. Dynamical systems with different values of a control parameter are distinct dynamical systems. Thus, a change in the values of the control parameter may lead to the appearance of other dynamical system behaviors as a result of bifurcations. This framework provides another manner for the implicit introduction of environmental variables into the system of interest as variables that control the system from outside. However, there exist situations in which the environmental variables cannot be reformulated in such a way; rather, they participate explicitly in determining the system's behaviors. As an example of a typical system, we studied the functional differentiation of cortical modules. Cell differentiation in embryos (e.g., the genesis of neurons) may represent another system.

20.2 Differentiation by Variational Principle

We studied a system with a constraint, which is supposed to reformulate the environmental variables and play a role that is distinct from that of the control parameters in self-organization. In this context, in contrast with the conventional self-organization described above, the functional elements (or components) are produced by such a constraint, which acts on the whole system [13, 15, 16]. In fact, we investigated the manner in which the functional elements emerge in a network system using two mathematical models: one addressed the emergence of neuron-like components from interacting discrete-time dynamical systems via the change of dynamical systems (model 1) [21], and the other addressed the emergence of hierarchical module-like components (model 2) [19, 23]. One may call this type of order formation self-organization of the second kind (SO2) [20]. It should be noted that a constraint in SO2 plays a different role from a constraint such as a boundary condition of a partial differential equation (PDE) in the development of systems. A boundary condition in PDE determines a type of pattern formation, which describes macroscopic orders in far-from-equilibrium systems. On the other hand, a constraint in SO2 determines the development of elements (components) of systems, hence referred to differentiation. To achieve the development of dynamical systems, a maximum transmission of information constraint was applied as a *variational* principle. Subsequently, we found a genesis of an excitable system for model 1, and a genesis of functional modules by spontaneous symmetry breaking for model 2. Let us briefly describe these models. In model 1, we studied a certain set of functions, using a single equation that possessed six parameters. The set of these parameters was viewed as genes. By introducing a discrete time step into these functions, we generated one-dimensional maps and their networks. Providing an external input to the network, we calculated time-dependent mutual information between input time series and the time series of each elementary dynamical system, and recorded its maximum value. By adopting a certain genetic algorithm, we observed the development of the network of dynamical systems, changing the set of parameters, and selected a dynamical system with maximum transmission of external information. Then, we observed an excitable dynamical system, which possesses specificity of neurons, as an asymptotic solution of the system development. In model 2, we treated functional differentiation of modules. Even in early periods of development of functional modules in the neocortex of mammalian brains, functional modules can be reorganized, depending on the environmental conditions. This implies the importance of environmental variables for the internal development of the brain. To investigate the manner in which such functional differentiation occurred, we considered two probabilistically identical modules that were coupled in a probabilistically uniform way, and then investigated the development of the network using a certain genetic algorithm. Then, we observed the following differentiation of modules. In one module, in-phase couplings were dominant, but other types of couplings also survived. In the other module, only in-phase couplings survived. The couplings between modules developed to be differentiated into in-phase couplings in one direction of the couplings and anti-phase couplings in the other direction. The number of couplings was also asymmetric. In these models, the environmental variables influenced the system in such a manner that the system development was operated under the constraint that represented the system adaptation to the environment. The long-term behaviors of the dynamics of order parameters observed, especially in model 2, indicated an extremely slow modulation of oscillations in a manner of chaotic itinerancy, [8, 17, 18], including intermittent transitions between synchronization and desynchronization, which were pointed out by Kozma and Freeman as a signature of field theory. A similar behavior has been observed in the rat hippocampus as second-order oscillations of the power of theta rhythms [9], and similar behaviors have also been observed in the dynamics of default mode networks [2]. Ontogenetic development of the network structure in the brains has also been attempted to describe in terms of exponentially expanding graph model by Freeman and Kozma [5]. However, in their model no mechanism to establish the reorganization of neural assemblies representing functional units seems to exist. Furthermore, because non-dissipated energy can be used for work, a part of which can be consumed to yield information carriers, our theory with a basis of optimization of information transmission may be compatible with a theory with a basis of optimization of free energy, e.g., Friston's theory [6], even though free energy is a concept in equilibrium systems, not in far-from-equilibrium systems. The computational results of our models suggest that these features of the dynamics observed in the brain are derived from a self-organization of the second kind, which may occur in cortical development, in contrast with the field theory of Kozma and Freeman, which seemingly fails to incorporate the development of dynamical systems that are directly driven by the environmental variables in the dynamics of the system itself.

Acknowledgments This work was partially supported by Grant-in-Aid, 26280093, JSPS.

References

1. Arbib M, Érdi P, Szentágothai J (1997) A bradford book. Neural Organization: structure, function and dynamics, MIT Press, Cambridge
2. Fox MD, Snyder AZ, Vincent JL, Corbetta M, van Essen D, Raichle ME (2005) The human brain is intrinsically organized into dynamic, anticorrelated functional networks. PNAS 102:9673–9678
3. Freeman WJ, Kozma R, Werbos PJ (2001) Biocomplexity: adaptive behavior in complex stochastic dynamical systems. Biosystem 59:109–123
4. Freeman WJ (2005) A field-theoretic to understanding scale-free neocortical dynamics. Biol Cybern 92:350–359
5. Freeman WJ, Kozma R, with appendix by Bollobas B, Riordan O, (2009) Scale-free cortical planar networks. Handbook of large-scale random networks. Springer, New York, pp 1–48
6. Friston K (2010) The free-energy principle: a unified brain theory? Nat Rev Neuro 11:127–138
7. Haken H (1977) Synergetics. Springer, Berlin
8. Kaneko K, Tsuda I (2003) Chaotic itinerancy. Chaos 13(3):926–936
9. Molter C, O'Neill J, Yamaguchi Y, Hirase H, Leinekugel X (2012) Rhythmic modulation of theta oscillations supports encoding of spatial and behavioral information in the rat hippocampus. Neuron 75(5):889–903
10. Nicolis G, Prigogine I (1977) Self-organization in nonequilibrium systems. Wiley, New York
11. Pias C (1946) Zeit der Kybernetik—Eine Einstimmung. In: Pias C (Hg), Cybernetics-Kybernetik: The Macy-Conferences, vol 1953, pp 9–41
12. Rasmussen M (2007) Attractivity and bifurcation for nonautonomous dynamical systems. Lecture Notes in Mathematics, Springer, Berlin
13. Rosen R (1991) Life Itself: a comprehensive inquiry into the nature, origin, and fabrication of life. Columbia University Press, New York
14. Scott AC (2002) Neuroscience: a mathematical primer. Springer, New York
15. Tsuda I (1984) A hermeneutic process of the brain. Prog Theor Phys Suppl 79:241–259
16. Tsuda I (2001) Toward an interpretation of dynamic neural activity in terms of chaotic dynamical systems. Behav Brain Sci 24:793–810 discussions 811–847
17. Tsuda I (2013) Chaotic itinerancy. Scholarpedia 8(1):4459
18. Tsuda I (2015) Chaotic itinerancy and its roles in cognitive neurodynamics. Curr Opin Neurobiol 31:67–71
19. Tsuda I, Yamaguti Y, Watanabe H (2015) Modeling the genesis of components in the networks of interacting units. In: Proceedings of ICCN 2013
20. Tsuda I, Yamaguti Y, Watanabe H (2015) Self-organization of the second kind. In: Proceedings of Waseda AICS symposium on new challenges in complex system science
21. Watanabe H, Ito T, Tsuda I (2011) Making a neuron model: a mathematical approach. In: 11th meeting of mechanisms of brain and mind. Niseko, Hokkaido, pp 11–13
22. Wilson HR, Cowan J (1972) Excitatory and inhibitory interactions in localized populations of model neurons. Biophys J 12:1–24
23. Yamaguti Y, Tsuda I (2015) Mathematical modeling for evolution of heterogeneous modules in the brain. Neural Netw 62(1755):3–10

Chapter 21
Commentary by Kazuyuki Aihara and Timothée Leleu

Overlapping and Non-interfering Waves of Bursts

Timothée Leleu and Kazuyuki Aihara

21.1 Introduction

Recordings of cortical activity suggest that waves of neural activity propagate over long distances on the cortical surface. Although the concept of waves is useful for describing the transfer of information between distant regions of the brain, the mechanisms that allow for numerous spatio-temporal patterns of activity to propagate in the neural substrate without interfering with one another are unclear. The description of scalar quantities alone, such as the firing rate of single neurons, is not sufficient to explain how neural processes that take place at the same time and in the same space can be distinguished. In order to solve the interference problem, it is necessary to consider the propagation of neural activity through mesoscopic volumes of the cortex.

Recent studies about the cortical microcircuitry have shown that, within an elementary volume extending 100 μm on the cortical surface, there are multiple interlaced subnetworks of densely connected neurons [8] as shown in Fig. 21.1a. These subnetworks cannot be spatially segregated and may be parts of distinct cell assemblies. Because there are distinct subnetworks within each elementary volume of the cortex, the neural activity at each position in space can be understood using a vector which components describe the activity of its subnetworks. The patterns of activity created within an elementary volume induce corresponding input patterns to neighboring ones through a "sea of weaker connections" [9] and, in turn, modify the patterns of activity in the network. The set of patterns of activity that can propagate depends on the synaptic connections within and between elementary volumes that result from learning and self-organization.

We argue that the widespread propagation of neural activity on the cortical surface depends significantly on the detailed microcircuitry among interlaced subnetworks

---

T. Leleu · K. Aihara
Institute of Industrial Science, The University of Tokyo, Tokyo, Japan

R. Kozma and W.J. Freeman, *Cognitive Phase Transitions in the Cerebral Cortex – Enhancing the Neuron Doctrine by Modeling Neural Fields*,
Studies in Systems, Decision and Control 39, DOI 10.1007/978-3-319-24406-8_21

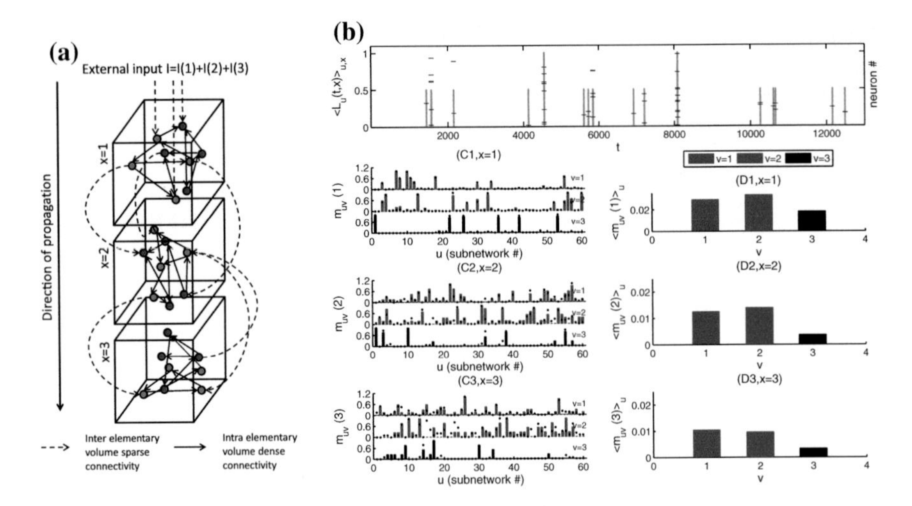

Fig. 21.1 **a** Microcircuitry of three elementary volumes in the cortex. **b** Bursts of the normalized averaged spiking activity $\langle L_u(t,x)\rangle_{u,x}$ at the time-step t and superimposed raster plot of spiking activity are shown by the *gray lines* and *short horizontal black lines*, respectively. **c1–c3** Conditional averaged avalanche sizes $m_{uv}(x)$ shown by *red*, *blue*, and *black bars* for avalanches triggered from the subnetworks in S_1, S_2, and S_3, respectively, when $\mathbf{I} = \mathbf{I_1} + \mathbf{I_2} + \mathbf{I_3}$. The *black dots* superimposed on the *red*, *blue*, and *black bars* show the averaged bursting activity $P_u(x)$ when $\mathbf{I} = \mathbf{I}_1$, $\mathbf{I} = \mathbf{I}_2$, or $\mathbf{I} = \mathbf{I}_3$, respectively. **d1–d3** Averaged avalanche size $\langle m_{uv}(x)\rangle_u$ in the elementary volume x

contained in each elementary volume. We generalize the concept of cellular automata, that describes the neural activity at a position using a scalar value [6], by showing that the patterns of activity of the subnetworks located in each elementary volume are represented by a vector. Some learned patterns of an elementary volume activate in turn other patterns in neighboring elementary volumes, whereas random patterns cannot spread.

21.2 Propagation of Patterns in Modular Networks

In order to illustrate the propagation of patterns in these modular networks, we consider three elementary volumes of the cortex along a single direction as shown in Fig. 21.1a. Each elementary volume is composed of N subnetworks. The weight matrix of connections between subnetworks within an elementary volume, indexed by its position x with $x \in \{1, 2, 3\}$, is noted $\Omega(x)$, and the weight matrices of connections from subnetworks of the elementary volume x_1 to the ones of the elementary volume x_2 are described by the matrices $A(x_1, x_2)$ with $(x_1, x_2) \in \{(1, 2), (2, 3), (2, 1), (3, 2)\}$. Moreover, we consider that the subnetworks of the first elementary volume receive an external input described by the vector $\mathbf{I} = \{\mathbf{I_u}\}_u$ where

I_u is the firing rate of Poisson distributed spikes inputted to the subnetwork u. The matrix $A(x_1, x_2)$ is sparse whereas the matrix $\Omega(x)$ has fewer zero elements.

The external input **I** results in a slow increase of the neuronal membrane potential in the first elementary volume until a spike is generated. This first spike can in turn induce a cascade of spiking activity, or burst, because of the dense connectivity within elementary volumes (see Fig. 21.1b). The spiking activity within elementary volumes is synchronous but sparse, and is well described by the concept of neuronal avalanches [5]. Each burst starts at a random time, when an external input destabilizes the current state, which is reminiscent of the random initiation time of amplitude modulated (AM) patterns [3] described in the framework of Freeman's Mass Action [2]. The results of simulations shown in Fig. 21.1 are based on the model of neuronal avalanches proposed in [1] for which the distribution of avalanches sizes in modular networks has been derived [7].

In these simulations, we ignore for the sake of simplicity the dynamics of synaptic currents and the delayed inhibition due to interneurons that usually participates in the conversion of spike trains into waves [3] and limit the scope of this study to the propagation of bursts. Thus, the nonlinear dynamics that results from the transformation of waves into pulses is not taken into account, nor is the related attractor dynamics. Note, however, that the nonlinearity due to the reset of membrane potential after spike generation is included in the analysis. The patterns of activity generated at each burst vary but can be characterized by a vector representing the temporal-averaged patterns. The impulse response of an elementary volume to an external input can be obtained by calculating the average number of neurons, noted m_{uv}, in the subnetwork u, $u \in \{1, \ldots, N\}$, that fire when an external input to the subnetwork v triggers an avalanche.

After learning, we consider that the connection strengths $\Omega(x)$ allow two spatio-temporal patterns, indexed by 1 and 2, to propagate from the first to the last elementary volume when the input patterns $\mathbf{I}(\mathbf{1})$ and $\mathbf{I}(\mathbf{2})$, respectively, are applied to the first elementary volume. These input patterns correspond to external signals to the cortex that are much weaker than the cortical response. The connection matrices $\Omega(x)$ that result from the Hebbian learning are given as $\Omega(x) = \sum_{v \in \{1,2\}} \omega_1(x, v)\omega_2(x, v)^T$ with $\omega_2(x, v) \propto A(x, x-1)\omega_1(x-1, v)$ and $\omega_1(1, v) \propto \mathbf{I}(\mathbf{v})$. For the sake of simplicity, the external input patterns $\mathbf{I}(\mathbf{v})$ and weight vectors $\omega_1(x, v)$ are set to be random vectors composed of βN components equal to zero ($\beta < 1$). Moreover, the matrices $A(x_1, x_2)$ are approximated to random permutation matrices.

We moreover take into account the role of another external input, noted $\mathbf{I}(\mathbf{3})$, that is chosen randomly and independently from the connection weights. $\mathbf{I}(\mathbf{3})$ represents the input patterns that are not associated to any learned spatio-temporal patterns. The averaged bursting activity is noted $P_u(x)$ and is defined as the average number of spikes emitted during a burst by the neurons of the subnetwork u in the elementary volume x. $P_u(x)$ is given as $P_u(x) = \frac{1}{N_a}\sum_a L_u(a, x)$ where $L_u(a, x)$ is the number of spikes emitted during the avalanche a in the subnetwork u of the elementary volume x and N_a is the total number of avalanches that occur during a trial.

When the external input **I** applied to the first subnetwork is the superposition of the patterns $\mathbf{I}(\mathbf{1})$ and $\mathbf{I}(\mathbf{2})$, both spatio-temporal patterns 1 and 2 propagate in the network. The superposition of the averaged activity patterns results in some interference between them: it is not possible to extract the two spatio-temporal patterns 1 and 2 separately using only the averaged firing rate. However, the bursts of spiking activity within elementary volumes allow encoding independently the two spatio-temporal patterns. In order to recover the two learned spatio-temporal patterns 1 and 2 from their superposition, we classify the bursts according to the subnetworks from which they were triggered. The set of subnetworks of the first elementary volume that receive spikes from the external inputs $\mathbf{I(1)}$, $\mathbf{I(2)}$, and $\mathbf{I(3)}$ are noted S_1, S_2, and S_3, respectively. The averaged avalanche size can be expressed as follows:

$$m_{uv}(x) = \langle L_{uv}(a, x)\rangle_a = \frac{1}{N_a(v)} \sum_a L_{uv}(a, x), \tag{21.1}$$

where $L_{uv}(a, x)$ is the number of spikes that are fired from the subnetwork u of the elementary volume x during avalanches a that are triggered from one of the subnetworks in the set S_v and $N_a(v)$ is the number of avalanches triggered from the subnetworks in the set S_v with $v \in \{1, 2, 3\}$. The averaged avalanche size $m_{uv}(x)$ can be obtained analytically and is given as follows [7]:

$$M(x) = (Id - \Omega(x))^{-1}, \tag{21.2}$$

where Id is the identity matrix and $M = \{m_{uv}\}_{uv}$.

Figure 21.1c1–c3 shows that the conditional averaged activity patterns $m_{uv}(x)$ observed when the external input $\mathbf{I} = \mathbf{I(1)} + \mathbf{I(2)} + \mathbf{I(3)}$ is applied to the first elementary volume are very similar to averaged activity patterns $P_u(x)$ observed when only the external input $\mathbf{I} = \mathbf{I(v)}$ is applied, $\forall v \in \{1, 2, 3\}$. When the external input is small, it can be shown that the averaged bursting activity, calculated without taking into account the subnetworks from which the avalanches start and denoted by $P_u(x)$, is a linear superposition of the averaged avalanche sizes $m_{uv}(x)$ such that $P_u(x) = \sum_v \alpha_v m_{uv}(x)$ with $\alpha_v > 0$. Classifying the avalanches according to the subnetworks from which they start allows recovering the averaged avalanche size $m_{uv}(x)$ from the spike train. Thus, synchronous bursts classified using their triggering subnetworks can be independent channels of communication.

In summary, the spatio-temporal patterns 1 and 2 propagate without interfering using time-division multiplexing of the synchronous bursts of activity that start from the subnetworks in S_1 and S_2, respectively. This is compatible with the concept of cinematographic frame [4]. Avalanches may correspond to frames and repeat 3 to 5 times per second. Each burst would likely take the form of AM patterns with a carrier frequency close to γ oscillations if the synaptic currents and delayed inhibition were taken into account.

Note that the spiking activity due to the external input $\mathbf{I(3)}$ propagates less than the other learned patterns (see Fig. 21.1d1–d3). Spatio-temporal patterns propagate only if the connections within and between elementary volumes allow these patterns to spread. The propagation of activity is thus dependent on the microcircuitry within elementary volumes. Although the precise connectivity of neurons within elementary volumes cannot be generally determined experimentally, the numerical simulations shown in Fig. 21.1 suggest that the detailed microcircuitry must be considered in order to account for the fact that certain patterns propagate on the cortical surface while others do not.

Interpreting the waves of activity on the cortical surface without knowing the details of the microcircuitry is one of the great dilemma of systems neuroscience. Our results suggest that the weight matrix of synaptic connections contained within each elementary volume can be of critical importance and determines the amplitude and direction of these waves. The modeling of bursts propagation and their associated waves can become increasingly complex as the number of overlapping patterns increases. In such a case, theoretical frameworks that allow keeping track of the creation and annihilation of propagating waves, inspired for example from Quantum Field Theory [10], may likely become very useful tools.

This research was funded by ImPACT Program of Council for Science, Technology and Innovation (Cabinet Office, Government of Japan).

References

1. Eurich CW, Herrmann JM, Ernst UA (2002) Finite-size effects of avalanche dynamics. Phys Rev E 66(6):066137
2. Freeman WJ (1975) Mass action in the nervous system. Academic Press, New York
3. Freeman WJ, Vitiello G (2009) Dissipative neurodynamics in perception forms cortical patterns that are stabilized by vortices. J Phys Conf Ser 174(1):012011 (IOP Publishing)
4. Freeman WJ, Kozma R (2015) (In this issue)
5. Friedman N, Ito S, Brinkman BA, Shimono M, DeVille RL, Dahmen KA, Butler TC (2012) Universal critical dynamics in high resolution neuronal avalanche data. Phys Rev Lett 108(20):208102
6. Kozma R, Puljic M, Balister P, Bollobás B, Freeman WJ (2005) Phase transitions in the neuropercolation model of neural populations with mixed local and non-local interactions. Biol Cybern 92(6):367–379
7. Leleu T, Aihara K (2015) Unambiguous reconstruction of network structure using avalanche dynamics. Phys Rev E 91(2):022804
8. Perin R, Berger TK, Markram H (2011) A synaptic organizing principle for cortical neuronal groups. Proc Natl Acad Sci 108(13):5419–5424
9. Song S, Sjostrom PJ, Reigl M, Nelson S, Chklovskii DB (2005) Highly nonrandom features of synaptic connectivity in local cortical circuits. PLoS Biol 3(3):e68
10. Vitiello G (2015) (In this issue)

Chapter 22
Commentary by Giuseppe Vitiello

Filling the Gap Between Neuronal Activity and Macroscopic Functional Brain Behavior

Giuseppe Vitiello

Abstract Two complementary approaches have been used to study brain and in general biological systems. In one of the approaches the brain-system is split into a large number of components, which are then studied in all their details. The problem of combining the data so accumulated in a working scheme able to account for the macroscopic observed functioning of the brain often is left unsolved since it is actually out of reach in this approach. Contradictory features often arise, indeed. For example, it is not clear how the high effectiveness and stability of some characterizing brain features may result from the random biomolecular activity of the brain component cells. A dilemma already pointed out by Lashley in neuroscience, and by Schrödinger in biology, but still waiting an answer. This first approach is the naturalistic approach. The other approach is the dynamical approach aiming to provide a comprehension of macroscopic features of the brain behavior on the basis of the data provided by the first approach. Both approaches appear thus to be necessary, although each one of them, separately considered, is not sufficient to account for the full understanding of brain functioning. A bridge between these approaches could be built following the strategy successfully used in the study of many-body condensed matter physics. In this direction moves the dissipative many-body model of brain, where the observed dynamic amplitude modulated (AM) assemblies of coherently oscillating neurons are described in the frame of the quantum field theory of spontaneously broken symmetry theories. Observations of scale free and critical phenomena in brain activity are also related to the coherent dynamics playing a crucial role in the dissipative model. A representation in terms of thermodynamic generalized Carnot-Rankine cycles is provided, which describes the process of formation of the coherent AM patterns as a transition from disordered, gas-like state of high entropy to liquid-like organized neuronal configurations of low entropy.

G. Vitiello
Dipartimento di Fisica "E.R. Caianiello", Universitá di Salerno, and INFN
Gruppo collegato di Salerno, 84084 Fisciano, SA, Italy

R. Kozma and W.J. Freeman, *Cognitive Phase Transitions in the Cerebral Cortex – Enhancing the Neuron Doctrine by Modeling Neural Fields*,
Studies in Systems, Decision and Control 39, DOI 10.1007/978-3-319-24406-8_22

22.1 The Brain Is Not a Stupid Star

It seems that Aristotele, in order to demysthicizing the stars, used to say that they behave in a stupid way, since they are passing always through the same point in their perennial motion, being in this way completely predictable. Symptom of intelligence is instead to change trajectory at any change of the initial conditions in order to achieve an intentional task. I do not know if Aristotele did really say or write such a statement. It is however clear that the brain does not behave as a stupid star. Not only. It does hit an intentional target with precision and determination, sometimes it misses it, it is true, but never the behavior and action of the brain in/on the world around it is void of some intentional plan. Even the wanderer moves around in the world searching for something, which is maybe not clear to him, but which at the end will open his eyes. Not many words are needed to say what we observe: the brain functional activity is extremely stable and at the same time extremely sensible to adapt its govern of the body to any change, even very tiny, of the environment. Of course, without such characterizing properties of the brain functioning, survival for us would be impossible.

Then the question immediately arises: how is it possible that myriads of complex components, such as a number of the order of 10^{11} neurons, 10^{15} synapsis, with their 10^{16} synaptic operations per second, each of them able to fire about 10 pulses per second, without considering glia cells, all of it in a bath of 90 % more numerous water molecules, each one carrying an electric dipole momentum oscillating under the influence of unavoidable quantum fluctuations, may generate such a precious stability of the brain overall functioning? It is evident that no answer can be babbled without the help of the enormous achievements of neuroscience, which explores the formidable biological and biochemical complexity of brain cells, even detecting the intricate nets connecting them and designing the maps of the locations of nervous cells apparently involved in specific tasks, such as vision, motor, etc. However, as it always happens in Science, also in neuroscience the growth of the knowledge of the properties of the elements under study is always also associated to the growth of the number of puzzling questions to be answered.

For example, how is it possible that the energy consumption per second by the whole brain activity is of the order of magnitude of 25 W, ridiculously small if compared with to the 1.5 MW necessary for the simulation of quite elementary tasks in one of the gigantic American or European Brain Projects? How the functioning of the brain continues to be efficiently stable, even when the integrity of the neuronal nets is highly compromised by the loss and/or malfunctioning of single neurons? The words of Schrödinger come to our mind. He was referring to biological systems in general, but they apply to brain also, which is doubtless one of the most complex living matter systems. Schrödinger was observing that the "regularities only in the average" [36] (p. 78), emerging from the *statistical mechanisms* is not enough to explain the *enigmatic biological stability* [36] (p. 47). He was stressing that the attempt to explain the biological functional stability in terms of the regularities of statistical origin would be the *classical physicist's expectation*' that *far from being trivial, is wrong* [36]

(p. 19). Of course, the data available at the time of Schrödinger observations were not so many and so detailed as they are today as a consequence of the enormous efforts and successes of molecular biology in general and neuroscience, for what concern the brain. However, it is a not deniable fact that the questions mentioned above are still there unanswered, lucidly synthesized in the forties by the Lashley dilemma [28]: "...Here is the dilemma. Nerve impulses are transmitted ...form cell to cell through definite intercellular connections. Yet, all behavior seems to be determined by masses of excitation...within general fields of activity, without regard to particular nerve cells... What sort of nervous organization might be capable of responding to a pattern of excitation without limited specialized path of conduction? The problem is almost universal in the activity of the nervous system."

About twenty years later, the situation was not much changed despite the enormous successes of neuroscience studies harvested in the two decades. In 1967, Ricciardi and Umezawa were writing [35]: "...it seems that a very few concrete results have been obtained, in the sense that the question *how the brain works out the information received from the outside, and which is the logic on which the operations performed by the brain are based* is still far from a satisfactory solution." The work of the "naïve physicist", as Schrödinger calls him [36] (p. 20), and I would say also of the "naïve neuroscientist", since we are talking of brain, apparently did not bring to concrete advancements on answering those basic questions. Ricciardi and Umezawa described further the situation by adding [35]: "...in the case of natural brains, it might be pure optimism to hope to determine the numerical values for the coupling coefficients and the thresholds of all neurons by means of anatomical or physiological methods. Moreover, as soon as one asks the question whether or not these models can be looked on as models of the true brain, many questions immediately arise. First of all, at which level should the brain be studied and described? In other words, is it essential to know the behavior in time of any single neuron in order to understand the behavior of natural brains? Probably the answer is negative. The behavior of any single neuron should not be significant for the functioning of the whole brain, otherwise higher and higher degree of malfunctioning should he observed, unless to assume the existence of "special" neurons, characterized by an exceptionally long half life: or to postulate a huge redundancy in the circuitry of the brain. However, up to our knowledge, there have been no evidences which show the existence of such "special" neurons, and to invoke the redundancy is not the best way to answer the question."

Moving along this line of thoughts, Ricciardi and Umezawa formulated the many-body model of brain [35] based on the study of "specific dynamical mechanisms (already known in physics of many degrees of freedom) which can satisfy the essential requirements of the observed functioning of the brain." The many-body model has been further developed in the following years [41, 42] so to include the crucial feature that consists in the fact that the brain is an *open* system, namely a system continuously exchanging energy and matter with the environment; it is a *dissipative* system. The dissipative model, of course, does not propose to forget or to bypass the efforts and the search made by neuroscientists in penetrating the secrets of the brain biological components, their distribution and organization. All of this is absolutely *necessary*.

The point is that the collection of data, sophisticate observations of the components properties, their connections in larger entities characterizes the *naturalistic* level of the research, which, as said, is necessary. However, it is *not sufficient* to the *comprehension* of the system at the level of their *collective* behavior. What is needed is the passage from the naturalistic description to the discovery and understanding of dynamical unifying laws. This is what happens in every sector of the scientific research, not only in neuroscience, of course.

Schrödinger [36], Herbert Fröhlich with his search of long range coherence in living matter [20, 21], Umezawa and Ricciardi [40], Karl Pribram with his holographic model [32, 33], and Walter Freeman with his direct focusing on mass action [6], each of them in their own historical time and field of interest, were thus addressing to the core of the problem, which is in fact in the Lashley words: "all behavior seems to be determined by masses of excitation...within general fields of activity, without regard to particular nerve cells". Here are the key words: "masses of excitation" and "fields of activity". Here is the turning point by which naturalism becomes Galilean science; in Schrödinger's words, in the study of living matter of crucial relevance is the distinction between the *two ways of producing orderliness* [36] (p. 80): ordering generated by the "statistical mechanisms" and ordering generated by "dynamical" interactions.

"...it needs no poetical imagination but only clear and sober scientific reflection to recognize that we are here obviously faced with events whose regular and lawful unfolding is guided by a *mechanism* entirely different from the *probability mechanism* of physics" [36] (p. 79).

In the Ricciardi and Umezawa model and in the subsequent dissipative many-body model of brain [41, 42], the *dynamics* is studied involving the *fields* and producing the *masses of excitation*, predicted by Lashley, with the help of modern quantum theories describing the intricate nonlinearity of the elementary component interactions.

In living matter, and in the brain as well, chemical changes, disassembling and replacements of biomolecules occur in a sort of turn over process in relatively short intervals of time (typically a couple of weeks) due to the metabolic activity. The high stability of brain functions and of long and medium life time memories could be then hardly explained in terms of specific, localized arrangements and ordering of biomolecules. On the contrary, as everywhere occurs in condensed matter, also in the brain long range neuronal correlations may account for memory recording as a response to external (or also endogenous) stimuli. Neurons, glia cells and all biologic entities are treated as classical object in the model. The many-body variable of quantum interest are the quantum fluctuations in the electrical dipole oscillatory modes, which are the ones underlying the long range dynamically generated correlations, sustaining the assembling and disassembling of large neuronal populations. Next Section is devoted to a very brief qualitative presentation of the main features of the dissipative model. In the subsequent Section, for the reader convenience we list the main results of the model as compared to the laboratory observations.

22.2 Far from the Equilibrium Systems

Living systems, including of course the brain, are far from the equilibrium open systems. This means that in their dynamical evolution they continuously undergo transitions through dynamical regimes, or *phases*, which are physically different from each other, since they manifest different phenomena of cellular organization and activity. In order to set up the canonical formalism, mathematics requires to double the system degrees of freedom to account for the environment or bath in which the system is embedded. In other words, one needs to define a sink in the environment where matter and energy released by the living system (source) go, and, vice-versa, identify a source in the environment for the matter and energy received by the living system (sink).

Since what is outgoing from the system is ingoing in the environment, and vice-versa, the mathematical description of such a *reversal* of out↔ in is easily achieved by reversing the sign of the time for one of the two parts (system/environment) with respect to the other one: assuming, as usual, positive sign of time for the system, for the environment one will use negative sign of time (time-reversal). Thus, since the outgoing flux from one of the two corresponds to the ingoing flux as seen by the other, receiving part, from such a perspective the two parts are one the 'time-reversed copy' of the other. Taking the living system as our reference, the environment is then described as the its "time-reversed copy".

In conclusion, one thus needs to include in the model the environment described as the system time-reversed copy, which is called the *Double* of the system [41, 42]. The reciprocal interactions system-environment have then the mathematical form of the system-Double interactions. This leads to consider the mathematical space of the phases, also called, in the physicist jargon, space of the unitarily inequivalent representations, where "unitarily inequivalent" has a precise mathematical definition (on which we do not insist) and is synonymous with "physically inequivalent", i.e. one of the representations is not reducible to another one without affecting the state of the system. There is thus the possibility to adopt a quantity whose value is different in different phases. Such a parameter is called order parameter. The value of the order parameter remains constant for changes of the system occurring within a given representation. In the dissipative many-body model of brain, different memories are recorded in different representations and the corresponding different values of the order parameter denote different specific memories. In this way memories are protected from *confusion* since the corresponding representations cannot be reduced one to the other one. The space of the representation is then recognized to be the space of the memories. The evolution of the brain is described by trajectories, with given initial conditions, in the space of the representation. These trajectories are found to be classical chaotic trajectories [9, 10, 24, 25, 29, 37, 39, 43]. Thus the model predicts quite different behaviors corresponding to slightly different initial conditions.

For what said above, in order to study the dynamical evolution of large populations of neurons, glia cells, biological units and the water bath in which they are embedded we need the mathematical framework characterized by the existence of infinitely many unitarily inequivalent representations. This is the framework of quantum field theory (QFT), indeed. Then, according to the model, in order to record one memory, we have to pick up one of the representations among the infinitely many ones. This choice is induced by an external (or endogenous) stimulus, the one associated to the memory it represents. Such a process is known as the mechanism of the spontaneous breakdown of symmetry. Choosing a representation is indeed forcing the system to sit in one of its available representations, all equally accessible. Before going on, I remark that we are working with *fields*, which is what we wanted according to the discussion in the previous Section. However, we are also working with *quantum* fields and this needs some clarification.

We already said that neurons, glia cells and other biological units are classical objects. Nevertheless, we do have in the brain quantum variables. These are the quantum vibrational modes of the electrical dipoles of the water molecules which constitute the bath where all biological cells and units are embedded. QFT is the available standard tool to treat such a background of quantum variables and fluctuations. The symmetry of such a molecular system is the rotational symmetry of the electrical dipole quantum degrees of freedom. Long range correlation modes are generated as a result of the breakdown of this symmetry.

The coexistence of the quantum dynamical level and the classical one occurs in the brain as it does in many instances in condensed matter physics. The question to be answered is the one of how to fill the gap between the level of the elementary electrical dipole fluctuations and biochemical random kinematics and the macroscopic level of the brain functional activity [5–7, 14, 15, 38]. In other words, the dynamical mechanism has to be found by which fluctuations at cellular and quantum level become negligible at the macroscopic level. In QFT this mechanism is provided by the dynamics generating coherent states. Indeed, for coherent states the ratio between fluctuations ΔN and the number N of elementary components is given by $\Delta N/N = 1/|\alpha|$, where $|\alpha|$ is a measure of the coherence. Thus ΔN is negligible with respect to N for high $|\alpha|$, which shows how crucial is the role of the coherent dynamics: coherence accounts for the brain functional macroscopic stability against its fluctuating microscopic and cellular activity. Since $N = |\alpha|^2$ we also see that we deal with large N, which requires in fact the use of fields. This explains why we need quantum fields.

The Goldstone theorem in QFT with spontaneous breakdown of symmetry predicts the existence of quanta (called the Nambu-Goldstone (NG) bosons or quanta). The (Bose-Einstein) condensation of the NG quanta in the system ground state is responsible for long range correlations among the system elementary constituents [2, 40]. Coherence is the result of the boson condensation of NG quanta. It is a manifestation of these long range correlations. They sustain the ephaptic [1, 22] neuronal interaction over large distances and generate the ordered, low entropy, amplitude modulated (AM) neuronal patterns observed to form during the cortex active state and dissolve as it returns after transmission to its high entropy state, ready to receive the trigger stimulus [3, 10, 17, 47]. The cortex thus undergoes state transitions through AM patterns forming in few ms, lasting for 80–120 ms, and reforming at rates in alpha and theta ranges (3–12 Hz) of EEG [6, 13, 14].

Topologically nontrivial phenomena have been observed to occur in the process of phase transitions, such as the formation of conic phase gradients, vortices, null spikes. A description of these phenomena is obtained by using the time dependent many-body Ginzburg-Landau equation [19] and a general picture is provided in terms of non-stationary thermodynamics. In such a scheme, the cortex goes through generalized Carnot-Rankine cycles [3, 17, 34, 47]. In the criticality domain the gas-like and the liquid-like phases coexist. In the building up of the liquid phase, oxidative metabolism provides the energy necessary to the cortex background high-density activity for its construction of knowledge by using selected information received through the perception channels. The cycle is then closed by returning to the gas-like phase of background low-density random firing of neurons with entropy increase (heat dissipation by cerebral circulation).

One remarkable aspect emerging in such a scenario is the interplay between linearity and nonlinearity: phase transitions and criticality domains typically occur in a nonlinear dynamical regime [17]. Within each representation, however, time evolution is controlled by linearity. The very same process of spontaneous breakdown of symmetry leading to the formation of coherent domains through the boson condensation process implies dynamical nonlinearity, which thus dominates phase transitions (critical phenomena where vortices and null spikes appear) [16]. However, within each phase linearity holds and activity is self-organized into a domain of superposition. One aspect of the relevance played by both these dynamical regimes has to do with the observation that wave density and pulse density coexist simultaneously in circular causality [7]. Pulse clouds built out of neural spike activity may be observed in linear response based experiments. However, their synchronization in AM domains shows log-log power density versus frequency distributions of the scale-free (self-similar fractal) cortical dynamics, which requires the nonlinear dynamics resting on spontaneous breakdown of symmetry and coherence.[1] Coherent states, indeed, may be shown to be isomorph to self-similar fractal properties [44–46]. The cell classical behavior thus emerges from the underlying coherent many-body dynamics [14, 16].

22.3 Conclusion

The brain, like other biological systems in general and condensed matter systems, is not the collection of its elementary and cellular components. If one wants to study how the brain extracts information from the many sensory receptors and constructs out of it meanings which constitute knowledge, two complementary approaches must be used. In one of them, the brain system is split into a large number of components which

[1]The phenomenon of decoherence, known to occur in quantum mechanics, does not affect QFT, as observed in coherent systems with very long life-time, such as superconductors, ferromagnets, crystals, which are described by coherent many-body dynamics. Coherent structures are observed to occur in condensed matter physics in a wide range of temperature, from quite high temperature (such as 3545 °C for diamond crystal melting) to very low temperature (such as −252 °C in superconductors).

are studied in all possible details and their distribution, arrangements and connections are described with great accuracy. This is the naturalistic level of the study, necessary but not sufficient in order to understand the brain functioning. Combining these data into a working scheme is the task of the second, complementary approach, the one of the study of the dynamics able to account of the brain-system behaviors and performances, a task generally out of reach of the first approach. The dissipative many-body model of brain, discussed in this paper, is an attempt to describe the dynamics which is object of study of the second approach. The discussion is aimed to shed some light on how to fill the gap between the basic component description of the naturalistic stage of brain studies and the observed macroscopic manifestations of the underlying dynamics at elementary level. The attempt is similar to the one successfully done in the study of condensed matter physics. This provides us the bridge between the naturalistic approach and the dynamical one. In some sense, the discussion here presented may be considered a contribution toward the solution of "the division problem of the world into parts to which an individual existence can be attributed" [23] (p. 1469), which Cassirer called the "ingenuous vision of the world" [4].

The observed dynamic amplitude modulated (AM) assemblies of coherently oscillating neurons are described in the frame of the quantum field theory of spontaneously broken symmetry theories [14, 15]. In the process of formation of the coherent AM patterns, the brain goes from disordered, gas-like, high entropy regime to liquid-like organized neuronal configurations of low entropy, and a representation in terms of thermodynamic generalized Carnot-Rankine cycles is proposed [3, 17, 18, 47]. The resulting thermodynamic model incorporates criticality and phase transitions. Long range correlations at the basic quantum level may sustain ephapsis in neuronal interactions generating AM patterns. The formation of topologically nontrivial structures, such as vortices, null spykes, phase cones observed in a criticality regime is described in the frame of the dissipative model. Nonlinearity and linearity, both manifest themselves in different dynamical regimes: nonlinearity in the phase transition critical processes, linearity in each of the phases where superposition of organized activity domains occur. Similarly, in the model wave density and pulse density may be shown to coexist in circular causality. Moreover, observed scale free, self-similar fractal distributions in log-log plots of energy density versus frequency [8, 11, 30, 31] are shown to be isomorph with the coherent state dynamics of the dissipative model.

The dissipative many-body model discussed in this paper is one of the possible attempts to study the brain functioning. Other approaches use Random Graph Theory [26, 27] and renormalization group formalism [12]. We do not discuss them in this paper. We observe that on the route from naturalism to a dynamical view in brain studies one might be inspired by the words of von Neumann [48]: "the mathematical or logical language truly used by the central nervous system is characterized by less logical and arithmetical depth than what we are normally used to. ...We require exquisite numerical precision over many logical steps to achieve what brains accomplish in very few short steps". Are we on the right track to discover and understand which ones are these *very few short steps*? Perhaps, together with Paul Werbos [49], we need to remain open-minded about all possible roads aiming to the understanding of brain functioning. The mathematical study of highly complex dynamical systems has shown in the past decades that we cannot avoid the use of field concepts and the richness of changes occurring in the symmetry properties in the course of the dynamical evolution of the system. Certainly, brains, although with their unique functional capabilities, are not isolated systems in the Universe and their study may even help us in recognizing the *inner life* of matter due to the quantum fluctuations of the vacuum. An integrated vision of Nature will then emerge [46].

We conclude by listing the main results of the dissipative model in agreement with laboratory observations. Below we closely follow the list in Ref. [16]. The results include:

- coexistence of physically distinct AM patterns in distinct frequency bands correlated with categories of conditioned stimuli,
- the rapid onset of AM patterns into (irreversible) sequences,
- very low energy required to excite AM correlated neuronal patterns,
- large diameters of the AM patterns with respect to the small sizes of the component neurons,
- duration, size and power of AM patterns are decreasing functions of their carrier wave number k,
- lack of invariance of AM patterns with invariant stimuli, but constancy with the unchanging meaning of the stimuli,
- self-similarity in brain background activity showing power-law distributions of power spectral densities derived from ECoGs data,
- heat dissipation at (almost) constant in time temperature,
- occurrence of near-zero down-spikes in phase transitions,
- the whole phenomenology of phase gradients and phase singularities in the vortices formation (criticality),
- the constancy of the phase field within the frames,
- the insurgence of a phase singularity associated with the abrupt decrease of the order parameter and the concomitant increase of spatial variance of the phase field,
- the occurrence of phase cones and random variation of sign (implosive and explosive) at the apex,
- the phase cone apices occur at random spatial locations,
- the apex is not initiated within frames, but between frames (during phase transitions),
- "classicality" (not derived as the classical limit, but as a dynamical output) of functionally self-regulated and self-organized background activity of the brain.

References

1. Arvanitaki A (1942) Effects evoked in an axon by the activity of a contiguous one. J Neurophysiol 5(2):89–108
2. Blasone M, Jizba P, Vitiello G (2011) Quantum field theory and its macroscopic manifestations. Imperial College Press, London
3. Capolupo A, Freeman WJ, Vitiello G (2013) Dissipation of dark energy by cortex in knowledge retrieval. Phys Life Rev 10:8594
4. Cassirer E (1968) Storia della filosofia moderna, vol 3. Il Saggiatore, Milano
5. Fingelkurts AA, Fingelkurts AA (2004) Making complexity simpler: multivariability and metastability in the brain. Int J Neurosci 114:843–862
6. Freeman WJ (1975/2004) Mass action in the nervous system. Academic Press, New York. http://sulcus.berkeley.edu/MANSWWW/MANSWWW.html
7. Freeman WJ (2001) How brains make up their minds. Columbia University Press, New York
8. Freeman WJ (2004) How and why brains create meaning from sensory information. Int J Bifurc Chaos 14:513–530
9. Freeman WJ (2009) Deep analysis of perception through dynamic structures that emerge in cortical activity from self-regulated noise. Cogn Neurodyn 3(1):105–116
10. Freeman WJ (2015) Mechanism and significance of global coherence in scalp EEG. Curr Opin Neurobiol 31:199–205
11. Freeman WJ, Breakspear M (2007) Scale-free neocortical dynamics. Scholarpedia 2(2):1357. http://www.scholarpedia.org/article/Scale-freeneocortical-dynamics
12. Freeman WJ, Cao Y (2008) Proposed renormalization group analysis of nonlinear brain dynamics at criticality, Chapter 27. In: Wang R, Gu F (eds) Advances in cognitive neurodynamics ICCN 2007. Springer, Heidelberg, pp 147–158
13. Freeman WJ, Quian Quiroga R (2013) Imaging brain function with EEG: advanced temporal and spatial imaging of electroencephalographic signals. Springer, New York. doi:10.1007/978-1-4614-4984-3. http://soma.berkeley.edu
14. Freeman WJ, Vitiello G (2006) Nonlinear brain dynamics as macroscopic manifestation of underlying many-body dynamics. Phys Life Rev 3:93118
15. Freeman WJ, Vitiello G (2008) Dissipation and spontaneous symmetry breaking in brain dynamics. J Phys A: Math Theor 41:304042. http://Select.iop.org.q-bio.NC/0701053v1
16. Freeman WJ, Vitiello G (2010) Vortices in brain waves. Int J Mod Phys B 24:326995
17. Freeman WJ, Capolupo A, Kozma R, Olivares del Campo A, Vitiello G (2015) Bessel functions in mass action modeling of memories and remembrances. Phys Lett A (in press)
18. Freeman WJ, Kozma R, Vitiello G (2012) Adaptation of the generalized Carnot cycle to describe thermodynamics of cerebral cortex. In: Proceedings of the IEEE world congress on computational intelligence WCCI/IJCNN. IEEE Press, Brisbane, pp 3229–3236
19. Freeman WJ, Livi R, Obinata M, Vitiello G (2012) Cortical phase transitions, non-equilibrium thermodynamics and the time-dependent Ginzburg-Landau equation. Int J Mod Phys B 26:1250035
20. Fröhlich H (1968) Long range coherence and energy storage in biological systems. Int J Quantum Chem 2:641–649
21. Fröhlich H (1977) Long range coherence in biological systems. La Riv del Nuovo Cim 7:399–418
22. Grundfest H (1959) Synaptic and ephaptic transmission. In: Fields J (ed) Handbook of physiology, vol 1, p 14798 [p 1902]
23. Haag R (1996) Understanding quantum field theory. Int J Mod Phys B 10:1469–1472
24. Kozma R, Freeman JW (2001) Chaotic resonance: methods and applications for robust classification of noisy and variable patterns. Int J Bifurc Chaos 10:2307–2322
25. Kozma R, Freeman WJ (2002) Classification of EEG patterns using nonlinear dynamics and identifying chaotic phase transitions. Neurocomputing 44:1107–1112
26. Kozma R, Puljic M (2015) Random graph theory and neuropercolation for modeling brain oscillations at criticality. Curr Opin Neurobiol 31:181–188

27. Kozma R, Puljic M, Balister P, Bollobs B, Freeman WJ (2005) Emergence of collective dynamics in the percolation model of neural populations: mixed model with local and non-local interactions. Biol Cybern 92:367–379
28. Lashley KS (1942) The problem of cerebral organization in vision. In: Cattell J (ed) Biological symposia VII, pp 301–322
29. Pessa E, Vitiello G (2003) Quantum noise, entanglement and chaos in the quantum field theory of mind/brain states. Mind Matter 1:59–79
30. Petermann T, Thiagarajan TA, Lebedev M, Nicoleli M, Chialvo DR, Plenz D (2009) Spontaneous cortical activity in awake monkeys composed of neuronal avalanches. Proc Natl Acad Sci 106(37):15921–15926
31. Plenz D, Thiagaran TC (2007) The organizing principles of neural avalanches: cell assemblies in the cortex. Trends Neurosci 30:10110
32. Pribram KH (1971) Languages of the brain. Prentice-Hall, Engelwood Cliffs
33. Pribram KH (1991) Brain and perception. Lawrence Erlbaum, Hillsdale
34. Rice SO (1950) Mathematical analysis of random noise—and appendices. Technical Publications Monograph B-1589. Bell Telephone Labs Inc, New York
35. Ricciardi LM, Umezawa H (1967) Brain and physics of many-body problems. Kybernetik 4:44–48. Reprint in: Globus GG, Pribram KH, Vitiello G (eds) Brain and being. John Benjamins, Amsterdam, pp 255–266, (2004)
36. Schrödinger E (1944) What is life? Cambridge University Press, Cambridge (1967 reprint)
37. Skarda CA, Freeman WJ (1987) How brains make chaos in order to make sense of the world. Brain Behav Sci 10:161–195
38. Stapp HP (2014) Mind, brain, and neuroscience, preprint March 5. University of California, Berkeley, California, Lawrence Berkeley Laboratory 94720
39. Tsuda I (2001) Towards an interpretation of dynamic neural activity in terms of chaotic dynamical systems. Behav Brain Sci 24:793–810
40. Umezawa H (1993) Advanced field theory: micro, macro and thermal concepts. American Institute of Physics, New York
41. Vitiello G (1995) Dissipation and memory capacity in the quantum brain model. Int J Mod Phys B 9:973–989
42. Vitiello G (2001) My double unveiled. John Benjamins, Amsterdam
43. Vitiello G (2004) Classical chaotic trajectories in quantum field theory. Int J Mod Phys B 18:785–792
44. Vitiello G (2009) Coherent states, fractals and brain waves. New Math Nat Comput 5:245–264
45. Vitiello G (2012) Fractals, coherent states and self-similarity induced noncommutative geometry. Phys Lett A 376:2527–2532
46. Vitiello G (2014) On the isomorphism between dissipative systems, fractal self-similarity and electrodynamics. Toward an integrated vision of nature. Systems 2:203–216
47. Vitiello G (2014) The use of many-body physics and thermodynamics to describe the dynamics of rhythmic generators in sensory cortices engaged in memory and learning. Curr Opin Neurobiol 31:712
48. von Neumann J (1958) The computer and the brain. Yale University Press, New Haven
49. Werbos PJ (2015) (Contribution in this book)

Epilogue

This volume aims at an ambitious goal, namely to reconcile field-theoretical approaches to large-scale cortical dynamics with the prevailing neuron doctrine in order to progress towards deciphering the brain codes. It is not the first time in the history of science that different views on a given research objective are considered contradictory rather than complementary. The brain, however, teaches us through its transient, metastable dynamics that opposing tendencies may coexist in a unified system. Indeed, large-scale synchronization gives way to rapid desynchronization in a fraction of a second, again and again, telling us that collective oscillations can collapse in a meaningful way to give way (momentarily) to extreme fragmentation of the brain dynamics reduced to the unit level. In this book we introduce this intriguing view of cognition and consciousness and describe mathematical approaches to address this complexity.

The basic tenet of our approach, following Von Neumann, is that brains use languages different from the dominant mathematics based on calculus and differential equations. Recent advances in network science and Random Graph Theory provide additional mathematical tools that can complement differential equations in describing very rapid transitions between highly organized and disorganized brain states and consider this process as a phase transition. The corresponding approach is called neuropercolation. The combination of neuropercolation and differential equations can provide a suitable tool to characterize the recently discovered brain dynamics as it progresses through spatial-temporal singularities.

The commentaries in the second half of our book provide an additional level of richness toward the goal of breaking the brain code. In the context of neuroscience theories of cognition and consciousness: Baars' contribution explores the hypothesis that the experimentally observed and theoretically described phase transitions in the cortex are manifestations of the conscious broadcast event predicted in his Global Workspace Theory (GWT) of consciousness and possibly related to thalamocortical projections. This fascinating idea consistent with our cinematic theory of cognition and it indicates avenues for future research endeavors. Bressler explores the possible mechanisms of interareal interaction in the cortex. The intuitive approach

R. Kozma and W.J. Freeman, *Cognitive Phase Transitions in the Cerebral Cortex – Enhancing the Neuron Doctrine by Modeling Neural Fields*, Studies in Systems, Decision and Control 39, DOI 10.1007/978-3-319-24406-8

assumes that the interactions are represented by the neuron-neuron connections. In an alternative approach, the significance of the population-to-population interaction is explored. This option is feasible, for example due to field effects. Bressler's convincing argument shows that the population-population interaction agrees more broadly with experimental data. Érdi and Somogyvári review the history of the neuron doctrine and various population models. They describe the Hudgkin-Huxley model as the dominant multi-compartmental modeling approach based on the neuron doctrine. On the other hand, population models have been useful as well, and they give as an example the neural network model using McCullough-Pitts neurons and Hebbian learning. This latter attempt can be viewed as the precursor of machine learning techniques and artificial neural networks; they provide powerful engineering tools, but do not come close to model the real life neural tissues, the authors contend. They introduce a current source density model based on their novel micro-electric imaging technique. Ohl reviews his decades-long fundamental research aiming at identifying large-scale coordinated activities in the cortex in the form of amplitude modulated (AM) patterns. His results indicate the co-existence of topographic (unit-based) and holographic (collective) encoding in the cortex. Moreover, he provides credible arguments describing how collective representations evolve from the fragmented, unit-based encoding as the result of learning effects.

The next group of commentaries concerns differential equations in brain modeling: Wright employs linear stochastic partial differential equations for modeling statio-temporal brain dynamics. He describes results using PDEs, which provide plausible interpretation of the gamma-bursts as he characterizes the repetitive frames in the cinematic theory of cognition. He applies mean field approaches to describe synchrony in his model and also explores the information storage capacity of the model systems. Liljenström describes upward and downward causation, as well as circular causation in his comprehensive model, which contains two parts. The neocortical part is described by Hodgkin-Huxley spiking neuron model, while the paleocortical model segment describes hippocampus and olfactory bulb using first-order ordinary differential equations with asymmetric sigmoid function following the line of Freeman K models. He describes noise effects and the emergence of macroscopic states from microscopic units. Brown and Hirsch introduce the mathematical tool of infinitesimal diffeomorphisms (ID) to convert the differential equations in Freeman K models into discrete mappings. The benefits of the approach are greater computational efficiency and improved accuracy. They introduce the ID equivalent of Freeman K0, KI, KII, and KIII sets and illustrate that the KII-ID and KIII-ID reproduces several important dynamic features of limit cycle and chaotic attractors. Brown provides a practical example of applying Freeman K models with infinitesimal diffeomorphisms using tennis education test bed. This is an appropriate initiative as the K sets can be used to model the intentional action-perception cycle. Preliminary results show significant benefit due to application of K sets in student tennis training examples.

The third group of commentaries describes novel theories for modeling collective dynamics and field effects: Werbos describes bottom-up models that are related to the neural dogma, while top-down models use field effects. He sees the corresponding roles of these two views for a comprehensive approach to brain functions. Considering the sudden switches from one frame to the other in the cinematic theory of cognition, Werbos proposes time-lagged recurrent neural networks (TLRNN) as an alternative approach instead of *phase transitions*. He emphasizes potential benefits of quantum theory of consciousness when addressing open questions of brain theories. Tsuda comments on the field theory approach and on phase transitions in the neuropercolation implementation of the hierarchy of Freeman K sets. In neuropercolation, as in any far form equilibrium system, control parameters are used to describe sudden changes in the system behavior. Tsuda's novel theoretical approach proposes a next level of complexity called *self-organization of the 2nd kind (SO2)*, when also the structure of the system changes as the result of parameter changes. Examples include the formation of the neuron structure and the ontogenetic development of biological systems, both emergent under the constraints of environmental effects. Aihara and Leleu address the role of hierarchy of brain models using neural subnetworks. Their model produces spatio-temporal patterns and they describe the role of the properties of the subnetworks in forming these patterns. Going beyond existing scalar-valued cellular automata models, they introduce vector models. Their approach includes learning, which is illustrated using a convincing computational example. Vitiello addresses the issue of the apparent contradiction between the neural level (microscopic) processing, on the one hand, and the brain level (macroscopic) behaviors, on the other hand. He proposes to marry these two extremes using many body approaches of dissipative quantum field theory. The model explains spontaneous symmetry breaking, scale-free behavior, as well as the repetitive phase transitions between disordered (gaseous) and ordered (liquid) phases. A list of the properties of the dissipative model is given, which are testable by laboratory experiments.

All these insights lead to the conclusion that understanding the brain codes clearly requires understanding its parts, including the neurons. However, this is not sufficient and the collective behavior of neural masses transcends the sum of the neural parts. This clearly includes field effects. Field effects lead to measurable electromagnetic signals which can be assessed non-invasively, using scalp EEG arrays. This potential is very valuable, and developments in the past decades provide ample evidence of the feasibility of non-invasive brain monitoring techniques.

Finally, let us introduce the following fanciful analogy (Anonymous Comments, 2014). Consider a fisherman and his cohort at a lake, who are observing the eddies and currents on the surface of the water, using buoyant flotation-sensors of minimal impact. They keep applying sophisticated tools to find statistical signs of the movements of the creatures (fish) in the lake without directly disturbing them. A different populous group of fishermen occupies another part of the lake. They are much less reserved, and they invade into the lake happily. They lower hooks, throw nets deep into the water, dive into it, and pull out wondrous creatures of many different shapes and sizes, dead and alive, to their great satisfaction. They proclaim: “This is the only true way to tell the story of the lake and its fish.” They laugh at the lonely fisherman and his cohort who carry on with their quiet observations of the life of the lake.

Illustration of the fishermen analogy to brain monitoring, based on anonymous critique (illustration by Vladimir Taytslin).

Illustration of the fishermen analogy to brain monitoring, based on anonymous critique (illustration by Vladimir Taytslin)

Index

R. Kozma and W.J. Freeman, *Cognitive Phase Transitions in the Cerebral Cortex – Enhancing the Neuron Doctrine by Modeling Neural Fields*,
Studies in Systems, Decision and Control 39, DOI 10.1007/978-3-319-24406-8

Printed in the United States
By Bookmasters